Copyright © 1991 by
THE NATIONAL COUNCIL OF TEACHERS OF MATHEMATICS, INC.
1906 Association Drive, Reston, Virginia 22091
All rights reserved

Library of Congress Cataloging-in-Publication Data:

Discrete mathematics across the curriculum, K–12 : 1991 yearbook /
Margaret J. Kenney, Christian R. Hirsch, editor.
    p.   cm.
Includes bibliographical references.
ISBN 0-87353-305-4
    1. Mathematics—Study and teaching. 2. Computer science—
Mathematics—Study and teaching.   I. Kenney, Margaret.
II. Hirsch, Christian R.
QA11.D56   1991   90-26358
510′.7—dc20   CIP

Printed in the United States of America

Contents

Preface

Change is the law of life. And those who
look only to the past or the present
are certain to miss the future.

John F. Kennedy, West Germany, 1963

The NCTM *Curriculum and Evaluation Standards for School Mathematics* is a framework for creating and implementing change in the teaching and learning of mathematics at all levels, K–12. The *Standards* is, in fact, a lens through which teachers, schools, and systems can view their present mathematics curriculum and see how it measures up against the needs of our increasingly technological society. One particular strand in the *Standards* that exemplifies the need for change is discrete mathematics.

This yearbook has been developed to give the mathematics education community some perceptions about discrete mathematics—why it is important, what it comprises at various grade levels, where it belongs in the curriculum—and some ideas about teaching it. The book is replete with practical ideas and suggestions for taking the first steps toward implementing discrete mathematics in the classroom. Simply put, discrete mathematics, as John Dossey explains in the first chapter, *is* the "math for our time."

A unifying thread, pervasive in many chapters, is that discrete mathematics incorporates many of the recommendations put forth in the *Standards*. A sampling includes the following:

- Discrete mathematics promotes the making of *mathematical connections*. For an example of this, explore chapter 16.

- Discrete mathematics provides a setting for *problem solving with real-world applications*. For an example, consider chapter 28.

- Discrete mathematics capitalizes on *technological settings*. For an example featuring calculators, see chapter 22. For an example focusing on computers, read chapter 14.

- Discrete mathematics fosters *critical thinking* and *mathematical reasoning*. For an example on mathematical reasoning, explore chapter 21.

vii

The yearbook is divided into nine sections. The first section builds a foundation with perspectives and issues. Chapter 1 sets the stage with background on the development of the theme and offers some instructional options. Chapter 2 sets some curriculum and pedagogical boundaries. The second section contains six chapters that describe discrete mathematics, activities, and problems for students in grades K–8. The third section offers an overview for the secondary school teacher and looks ahead to the college level. Each of the next five sections focuses on a significant topic from discrete mathematics: graph theory; matrices; counting methods; recursion, iteration, and induction; and algorithms. The ninth section includes chapters detailing materials for secondary school students.

The creation of a yearbook is an exciting process that brings together many capable and dedicated professionals. Sincerest thanks are given to all who played a role in the development of this book. This salute encompasses those who answered the call and submitted preliminary drafts as well as those who became chapter authors. The authors, whose combined efforts, prompt attention to all the production details, and thoughtful, insightful work, are truly deserving of recognition and praise. I am also most grateful to the NCTM headquarters staff, who have a gift for turning a pile of manuscripts into an impressive book. Grateful appreciation is given to Cynthia Rosso, Charles Clements, Jo Handerson, and others in Reston for their expert editorial and production assistance.

Very special thanks go to the advisory panel for a task well done:

Dwayne Cameron	Old Rochester Regional High School, Mattapoisett, Massachusetts
Timothy V. Craine	Trinity College, Hartford, Connecticut
Claire Z. Graham	Framingham State College, Framingham, Massachusetts
Christian R. Hirsch	Western Michigan University, Kalamazoo, Michigan
James T. Sandefur, Jr.	Georgetown University, Washington, D.C.

In particular, Chris Hirsch, the general editor for yearbooks, must be singled out for sharing his expertise.

The content under the umbrella of discrete mathematics, as cited by many of the chapter authors, is exceptional because it is motivational, it is appealing, and it helps to make mathematics come alive for students. Our intent in this yearbook is to share this observation, and we trust that many more students will come to view mathematics as a vibrant, continually evolving, useful subject.

MARGARET J. KENNEY
1991 Yearbook Editor

Discrete Mathematics: The Math for Our Time

John A. Dossey

T HE branch of mathematics known as *discrete mathematics* has rapidly grown in prominence in the past decade. This growth is due in large part to the many applications of its principles in business and to its close ties to computer science. The theorems and problem-solving strategies central to discrete mathematics, combined with the increased computational power of computers, have opened whole new areas of investigation and applications. However, average citizens, and many mathematics teachers, have never heard of discrete mathematics. What is it?

The dictionary defines *discrete* as "distinct from others; separate; consisting of distinct parts; discontinuous" (Thorndike and Barnhart 1983). Discrete mathematics, then, potentially involves the study of objects and ideas that can be divided into "separate" or "discontinuous" parts. Thus discrete mathematics can be contrasted with the classical notion of continuous mathematics, which is the mathematics underlying most of algebra and calculus. These two topics typically use the real or complex numbers as a domain for their functions. Discrete mathematics, by contrast, is needed for the investigation of settings where functions are defined on discrete, or finite, sets of numbers, such as the positive integers.

Continuous mathematics is well suited to situations whose main objective is the measurement of a quantity. In discrete mathematics settings, the focus is on determining a count. Although some would place the two branches of continuous and discrete mathematics in head-to-head competition, reality shows that both approaches complement each other in real-world applications. For example, discrete approaches give approximations to the size of some measurements, whereas continuous methods allow the establishment of some bounds for the number of steps in computing algorithms that are finite in nature.

Discrete mathematics problems can be classified in three broad categories. The first category, *existence problems,* deals with whether a given problem has a solution or not. The second category, *counting problems,* investigates how many solutions may exist for problems with known solutions. A third category, *optimization problems,* focuses on finding a best solution to a particular problem.

As an example of an existence problem, consider the construction of test forms for use in a state assessment program. Suppose there are seven basic groups of items. However, the time allowed for testing makes it impossible for each child to work on more than three groups of items. A psychometrician working on the project says that if it is possible to combine the groups in test booklets three at a time so that any two groups appear together in a test booklet exactly once and so that any two booklets have only one group in common, she can use statistics to project student scores on the item groups they did not work on. The budget officer for the assessment project says that the project cannot afford to print more than a total of twenty-one groups of items, no matter how they are joined in actual test booklets. Can the assessment program meet these criteria? If the groups are numbered 1, 2, 3, 4, 5, 6, and 7, the following assignment of blocks to test booklets will satisfy the conditions:

$$[1, 3, 7], [1, 2, 4], [2, 3, 5], [3, 4, 6], [4, 5, 7], [1, 5, 6], [2, 6, 7]$$

This listing shows the existence of a solution to the problem. This listing is also optimal, since fewer than seven booklets would not suffice for the conditions listed (Ryser 1963). The full solution to this existence problem involves counting and has strong ties to finite projective planes (Hall 1967).

Counting problems can take many forms. One commonly seen example to which every student can relate is the availability of telephone numbers. An individual telephone number consists of a three-digit area code and a seven-digit local number. An area code cannot begin with 0 or 1 and must have a 0 or a 1 as its middle digit. A local number is a sequence of seven digits that cannot have either a 0 or a 1 as its first digit. How many telephone numbers, area code and local number, are possible? There are eight choices for the first digit, two choices for the second, ten for the third, and so on. A straightforward application of the multiplication principle gives us the following result:

$$8 \times 2 \times 10 \times 8 \times 10 \times 10 \times 10 \times 10 \times 10 \times 10,$$

or 1 280 000 000 possibilities.

Optimization problems are among the most interesting encountered. Consider the problem of deciding which of a group of university experiments can be taken on a Discovery shuttle mission. After the NASA experiments are loaded, a maximum of 700 kilograms of payload is available for other experiments. Universities apply for shuttle space by submitting a proposal along with the space and mass, measured in kilograms, required. A team of experts rates the proposals on a scale of 1 (low) to 10 (high) on scientific importance. The data for the experiments are shown in figure 1.1.

Experiment	Mass in kg	Scientific rating
1. Cosmic rays	80	6
2. Weightless vines	25	3
3. Binary stars	224	4
4. Mice tumors	65	8
5. Space dust	127	7
6. Solar power	188	7
7. Relativity	104	5
8. Seed viability	7	5
9. Sun spots	92	2
10. Speed of light	324	8
11. Cloud patterns	36	6
12. Yeast	22	4

Fig. 1.1

If it is decided to choose the experiments so that the total of the scientific ratings is as large as possible and the mass total does not exceed 700 kilograms, how should the experiments be selected? To check out all possibilities for loading different subsets of these experiments on the flight, we would need to consider 2^{12} different combinations—all possible subsets of the proposals. Such problems are known as **knapsack** problems. To date no efficient method has been found to solve such problems (Dossey et al. 1987).

Problems like the three illustrated and the development of algorithms for their solutions are the core of discrete mathematics. Discrete mathematics can be characterized as the mathematics of finite situations that require the establishment of the existence of a solution, the number of possible alternatives, or the identification of the best solution for a specified problem.

THE GENESIS OF THE DISCRETE MATHEMATICS CURRICULUM

Discrete mathematics grew out of the mathematical sciences' response to the need for a better understanding of the combinatorial bases of the mathematics used in the development of efficient computer algorithms, the creation of new approaches to operations research problems, and the study of the heuristics underlying the approaches to such problems. The existence of discrete mathematics as a separate area of study began in the late 1960s. By the early 1970s a number of influential texts appeared at the upper undergraduate level (Bondy and Murty 1976, Liu 1968, Roberts 1976, Stanat and McAllister 1977, and Tucker 1980).

The textbooks were soon followed by recommendations for the inclusion of programs of study in the college curriculum for mathematics majors. These calls began with the Mathematical Association of America's *Recommendations for a General Mathematical Sciences Program* (Tucker 1981). Both of the recommended undergraduate sequences suggested for the mathematics major included work in discrete mathematics. The course outlines called for content from graph theory (properties, trees, graph coverings, circuits, and graph models) and combinatorics (counting principles, permutations and combinations, inclusion/exclusion, and recurrence relations). This agenda for change gained momentum through the recommendations of a conference on the mathematics of the first two years of the undergraduate experience held at Williams College in the summer of 1982 (Ralston and Young 1983). The recommendations underscored the need and further specified the content of offerings in discrete mathematics for the mathematics major. Connections with the secondary school curriculum were also considered (Maurer 1983). These connections were quickly followed by additional recommendations on the need to develop ideas of discrete mathematics earlier in the mathematics curriculum. The Conference Board of the Mathematical Sciences report *The Mathematical Sciences Curriculum K–12: What Is Still Fundamental and What Is Not* (1983) joined with the NCTM report *Computing and Mathematics: The Impact on Secondary School Curricula* (Fey 1984) in calling for incorporating discrete mathematics in the school mathematics curriculum.

The support of these reports and other suggestions for reform in the school mathematics curriculum (Hirsch 1985, Ralston 1985, Sandefur 1985) provided the basis for a number of NSF-sponsored teacher enhancement projects as well as other special activities to bring teachers up to date in discrete mathematics and its applications. This growth in teachers' knowledge, the appearance of discrete topics in commercial textbooks, and the publication of articles dealing with discrete mathematics in the *Mathematics Teacher* fostered the case for discrete mathematics. The inclusion of discrete mathematics as one of the secondary-level standards in the NCTM's *Curriculum and Evaluation Standards for School Mathematics* (1989) furnished the final impetus for serious consideration of the topic in all school mathematics programs.

The statement on discrete mathematics in the *Standards* reads, in part, as follows (NCTM 1989, p. 176):

In grades 9–12, the mathematics curriculum should include topics from discrete mathematics so that all students can—

- *represent problem situations using discrete structures such as finite graphs, matrices, sequences, and recurrence relations;*
- *represent and analyze finite graphs using matrices;*
- *develop and analyze algorithms;*

- *solve enumeration and finite probability problems;*

and so that, in addition, college-intending students can—
- *represent and solve problems using linear programming and difference equations;*
- *investigate problem situations that arise in connection with computer validation and the application of algorithms.*

Focus

As we move toward the twenty-first century, information and its communication have become at least as important as the production of material goods. Whereas the physical or material world is most often modeled by continuous mathematics, that is, the calculus and prerequisite ideas from algebra, geometry, and trigonometry, the nonmaterial world of information processing requires the use of discrete (discontinuous) mathematics. Computer technology, too, wields an ever-increasing influence on how mathematics is created and used. Computers are essentially finite, discrete machines, and thus topics from discrete mathematics are essential to solving problems using computer methods. In light of these facts it is crucial that all students have experiences with the concepts and methods of discrete mathematics.

This standard provides a set of goals for discrete mathematics for the high school years. Although it recommends that discrete mathematics topics be integrated throughout the high school curriculum, it does not suggest a manner for accomplishing the desired outcomes. During the summer of 1989, the Exxon Foundation gave the National Council of Teachers of Mathematics financial support for a group of teachers familiar with discrete mathematics to develop an outline for integrating these topics into the middle and secondary school curriculum (NCTM 1990). The group's recommendations focused on discrete topics from the areas of algorithmic thinking, critical reasoning, counting, discrete probability, matrices, graph theory, and iteration and recursion. The diverse nature of these topics permits multiple opportunities for their integration across the curriculum. Discrete topics often provide easy avenues of approach to significant mathematics for many students because of their intimate relationship with the natural numbers. However, this seemingly simple aspect also masks the deep nature of many of the questions to which discrete mathematics is often applied.

The NCTM report gives suggestions for integrating topics from each of the seven areas into the curriculum (*a*) at levels prior to the formal study of algebra, (*b*) in conjunction with algebraic topics, and (*c*) in conjunction with geometric topics. Many of the topics can be approached meaningfully at much lower grade levels. Determining the shortest path between two towns (vertices) on a map (graph) can be explored fruitfully at many grade levels starting in the second grade; the complexity of the map determines the grade level. The analysis of the recursive nature of a pattern can be explored across the grades, from simple extensions of known cases to the determination of a closed-form functional representation for the pattern. The table in figure 1.2 gives some sample suggestions for activities appropriate for students at each of the three secondary levels mentioned. The student-level outcomes described represent the types of skills and conceptual understandings the study of discrete mathematics should bring to the student by the secondary school level.

Topic	Prior to Algebra	In Conjunction with Algebra	In Conjunction with Geometry
Algorithmic thinking	Writes a detailed list of instructions for placing a long-distance telephone call.	Employs the nesting method for evaluating a polynomial on a hand calculator.	Analyzes the geometry of root-finding algorithms, such as the secant or bisection methods.
Critical thinking	Applies the meaning of "and," "or," "not," and "if-then" in classification activities.	Uses logical analysis in considering the possible solutions to quadratic inequalities and absolute-value settings.	Recognizes patterns in tables of data tied to spatial relationships: Euler's formula, number of spanning trees for a finite graph. . . .
Counting	Employs matrices to establish a model for a counting situation, perhaps an inventory matrix of types of clothing by color.	Applies tree diagrams to solve counting problems, such as: How many ways can one arrange a red, a blue, and a yellow pillow left to right on a couch?	Uses two different ways of counting the same set of objects and uses the counts to establish a relationship; for example, uses the number of faces at a vertex of a dodecahedron to find the number of edges.
Probability	Determines if a game employing a given sample space for possible outcomes is a fair game.	Uses the binomial expansion to calculate the probability of a binomial event.	Simulates the area of a given geometric figure using a Monte Carlo approach, that is, uses a simulation to find the area of
Matrices	Uses inventory matrices to make decisions on ordering inventory for a store.	Applies row-and-column operations in a manner similar to Gaussian elimination to find the inverse of a matrix.	Finds powers of a matrix to determine the number of paths between two vertices in a graph when given the adjacency matrix for the graph.

Graph theory	Finds the latest starting date for a stage in a process represented by a PERT chart.	Uses tree transversal methods to analyze the order of operations in an algebraic expression.	Employs graph-coloring methods to minimize the potential conflicts in scheduling meetings for a number of committees with joint memberships.
Iteration and recursion	Writes the first seven terms of sequences, such as: 2, 4, 7, . . ., $n + 2^{n-1}, \ldots$	Finds the sum of the first n terms of a geometric sequence with first term a and common ratio d, using recursive methods.	Solves geometric puzzles, such as: "Into how many regions is a circle divided by n secant lines?"

Fig. 1.2. Sample tasks

The project report also outlines a semester-long course designed for students who have completed the equivalent of second-year algebra. This outline emphasizes developing students' abilities to model problems, apply technology, use and analyze algorithms, think recursively, use mathematics to make decisions, and use inductive methods to solve discrete problems. The major features of the outline follow:

I. Social decision making (4–5 weeks)
  A. Fair division problems
  B. Election procedures (ranking methods, Arrow's fairness criteria for ranking and his impossibility theorem, voting paradoxes, and weighted voting and power indexes)
  C. Apportionment (methods of apportionment, apportionment paradoxes, Balinski and Young impossibility theorem)

II. Graph theory (3–4 weeks)
  A. Program evaluation and review technique (PERT) methods
  B. Minimal spanning trees (Prim's and Kruskal's algorithms)
  C. Structure of a graph (basic concepts, representations [diagrams, adjacency matrix, adjacency lists], breadth and depth—first searches)
  D. Circuits and paths (Euler circuit algorithm, Hamiltonian circuits and paths, traveling salesman problem, Dijkstra's algorithm for shortest paths)
  E. Graph coloring
  F. Structure of trees

III. Counting techniques (4–5 weeks)
  A. Logic, sets, and Venn diagrams (disjunction and union, conjunction and intersection, negation and complement, principle of inclusion-exclusion)

    B. Addition and multiplication principles
    C. Permutations and combinations
    D. Pascal's triangle
    E. Discrete probability and applications (mutually exclusive events and addition rule, independent events and the multiplication rule, conditional probabilities, and expected value)

  IV. Matrix models (1–2 weeks)
    A. Markov chains
    B. Leslie model for population distributions
    C. Leontief input-output model of an economy

  V. Mathematics of iteration (3–4 weeks)
    A. Iterating first-order recurrence relations
    B. Applications of iteration (arithmetic and geometric sequences, exponential growth, mathematics of finance, and population dynamics)
    C. Finding the closed form of a first-order linear recurrence relation
    D. Iterating second-order difference equations (Fibonacci sequence)

In addition to these topics, the outline provides references and gives suggestions for optional topics to expand the course. This set of recommendations parallels the work of the Mathematical Association of America's Committee on Discrete Mathematics in the First Two Years (1986). Both reports offer guidelines to schools, developers of curriculum materials, and teachers on possible paths to follow in establishing discrete mathematics as a part of the school mathematics program.

CONCLUSION

Discrete mathematics allows students to explore unique problem situations that are not directly approachable through writing an equation or applying a common formula. Students are often required to visualize the situation through developing a model or another form of representation. Other settings call for analyzing special cases or developing a solution by considering a simpler problem involving fewer cases. The theory does not require learning a large number of definitions and theorems but does require a sharp and inquisitive mind. Careful development of the content of discrete mathematics, building its ties to the current school program, and strengthening the connections between continuous and discrete approaches to topics can do much to give students the opportunities intended by the *Standards* and needed by young adults preparing to enter the world of work in the twenty-first century.

REFERENCES

Bondy, John A., and U.S.R. Murty. *Graph Theory with Applications.* New York: American Elsevier, 1976.

Conference Board of the Mathematical Sciences. *The Mathematical Sciences Curriculum K–12: What Is Still Fundamental and What Is Not.* Report to the NSB Commission on Precollege Education in Mathematics, Science, and Technology. Washington, D.C.: CBMS, 1983.

Dossey, John A., Albert Otto, Lawrence Spence, and Charles Vanden Eynden. *Discrete Mathematics.* Glenview, Ill.: Scott, Foresman & Co., 1987.

Fey, James T., ed. *Computing and Mathematics: The Impact on Secondary School Curricula.* Reston, Va.: National Council of Teachers of Mathematics, 1984.

Hall, Marshall, Jr. *Combinatorial Theory.* Waltham, Mass.: Blaisdell, 1967.

Hirsch, Christian R., ed. *The Secondary School Mathematics Curriculum.* 1985 Yearbook of the National Council of Teachers of Mathematics. Reston, Va.: The Council, 1985.

Liu, Chung, L. *Introduction to Combinatorial Mathematics.* New York: McGraw-Hill, 1968.

Mathematical Association of America. *Report of the Committee on Discrete Mathematics in the First Two Years.* Washington, D.C.: MAA, 1986.

Maurer, Stephen. "The Effects of a New College Curriculum on High School Mathematics." In *The Future of College Mathematics: Proceedings of a Conference/Workshop on the First Two Years of College Mathematics,* edited by Anthony Ralston and Gail S. Young. New York: Springer-Verlag, 1983.

National Council of Teachers of Mathematics. *Curriculum and Evaluation Standards for School Mathematics.* Reston, Va.: The Council, 1989.

———. *Discrete Mathematics and the Secondary Mathematics Curriculum.* Reston, Va.: The Council, 1990.

Ralston, Anthony. "The Really New College Mathematics and Its Impact on the High School Curriculum." In *The Secondary School Mathematics Curriculum,* 1985 Yearbook of the National Council of Teachers of Mathematics, edited by Christian R. Hirsch. Reston, Va.: The Council, 1985.

Ralston, Anthony, and Gail S. Young, eds. *The Future of College Mathematics: Proceedings of a Conference/Workshop on the First Two Years of College Mathematics.* New York: Springer-Verlag, 1983.

Roberts, Fred. *Discrete Mathematical Models.* Englewood Cliffs, N.J.: Prentice Hall, 1976.

Ryser, Herbert J. *Combinatorial Mathematics.* Carus Mathematical Monograph No. 14. Washington, D.C.: Mathematical Association of America, 1963.

Sandefur, James T., Jr. "Discrete Mathematics: The Mathematics We All Need." In *The Secondary School Mathematics Curriculum,* 1985 Yearbook of the National Council of Teachers of Mathematics, edited by Christian R. Hirsch. Reston, Va.: The Council, 1985.

Stanat, Donald, and David McAllister. *Discrete Mathematics in Computer Science.* Englewood Cliffs, N.J.: Prentice Hall, 1977.

Tucker, Alan. *Applied Combinatorics.* New York: John Wiley & Sons, 1980.

Tucker, Alan, ed. *Recommendations for a General Mathematical Sciences Program.* Report of the Mathematical Association of America's Committee on the Undergraduate Program in Mathematics. Washington, D.C.: MAA, 1981.

Thorndike, E. L., and Clarence L. Barnhart. *Scott, Foresman Advanced Dictionary.* Glenview, Ill.: Scott, Foresman & Co., 1983.

2

A Cautionary Note

Anthony D. Gardiner

DISCRETE mathematics is not new. Many of its fundamental ideas and techniques emerged in the eighteenth century—especially with Euler. But these were not recognized as a separate branch of mathematics until relatively recently. It is only during the last forty years that discrete mathematics has gained its independence and has become increasingly important both in mathematics and in everyday life. This is partly due to developments within mathematics itself, but the decisive factor has been the meteoric rise of computers. Indeed, one definition of discrete mathematics might be "those parts of mathematics that are *most obviously relevant* to computing."

The interaction of discrete mathematics and computers has made possible powerful new applications, has focused attention on new kinds of problems, and has forced us to look at traditional mathematics in new ways. Dramatic as these changes have been, and though they warrant our serious consideration, they should not be used to justify an overhasty rejection of traditional mathematics in favour of what we imagine to be more relevant material.

> Computer science penetrates classical mathematics putting old questions in a new perspective. . . . This new perspective on . . . classical mathematics is of course a challenge to mathematics educators. The introduction of computers (at whatever level) is only a very partial answer . . . it will take much more work—experimental and theoretical—before the contours of the answer will be clear. (Lovász 1988, p. 69)

> Teaching should not be subjected to sudden swings under the capricious blast of ephemeral fads. (Poincaré 1914, p. 151)

Few classical mathematicians were closer to the modern spirit than Poincaré. Few modern mathematicians have contributed more impressively to recent developments than Lovász. Yet when it comes to revising mathematics for ordinary students, both go out of their way to urge caution. Before looking more closely at the question of discrete mathematics at the school level, I shall try to explain why we should take their warnings seriously.

Education is a long period of *preparation*. Even when one knows exactly what one's students are preparing for—to speak French, to play Mozart, to survive in a drug-ridden society, to raise a family successfully, and so on—the appropriate preparation at the school level is likely to be only loosely related to these ultimate objectives. Supposing we knew exactly what mathematical skills future high school graduates should have, it would still not be clear exactly what preparatory experiences would help, rather than hinder, their eventual attainment.

We must also bear in mind just how fast things have changed in the last twenty years. In 1973 the most perceptive mathematician and educator of modern times could write, quite correctly, that "modern mathematics is distinguished from the old by the stress on the conceptual component as opposed to algorithms" (Freudenthal 1973, p. 44). Freudenthal went on to point out that conceptual and algorithmic mathematics should not be seen as two opposing camps and that the relationship between them is an organic one. Nevertheless, the emphasis was already changing at the time he was writing and is now markedly different.

> The design and analysis of algorithms and the study of algorithmic solvability uses deeper and deeper tools from classical structural mathematics . . . and an algorithmic perspective has a more and more profound effect on the whole framework of many fields of classical mathematics. (Lovász 1988, pp. 67–68)

And the emphasis will go on changing. Indeed, other areas of mathematics are already beginning to rival discrete mathematics in its own backyard—in the design and analysis of algorithms and computational mathematics in general. After discussing a number of lovely examples, Lovász concludes:

> What does this imply for mathematics education? Whatever it implies should be regarded with the utmost caution and moderation. I feel that math education should follow what happens in math research, at least to a certain extent, in particular those (rare) developments there that fundamentally change the whole framework of the subject. Algorithmic mathematics is one of these. However, the range of the penetration of an algorithmic perspective in classical mathematics is not yet clear at all. . . . Our experience with "New Math" . . . warns us that drastic changes may be disastrous—even if the framework is well established in research and college mathematics. (Lovász 1988, p. 75)

School mathematics has to evolve in a controlled fashion. It must never allow itself to be carried away by the latest "ephemeral fad," even if this means continuing for a time to teach material that *seems* old-fashioned:

> It is hard for the teacher to teach material which does not completely satisfy him; but the satisfaction of the teacher is not the sole object of teaching; one's primary concern should be for the pupil's own mental world and how one wants it to develop. (Poincaré 1914, pp. 134–35)

Thus, although we should continually look for ways in which school mathematics can better reflect changes in mathematics and in everyday life, we must always remember two points: first, that the most appropriate *introduction* may look quite different from the developments that originally motivated its inclusion at the school level, and second, that in choosing appropriate material, we must concentrate on things that most youngsters and their teachers can handle in a meaningful way on their level.

IMPLICATIONS FOR DISCRETE MATHEMATICS AT THE SCHOOL LEVEL

What does all this have to tell us about the possible introduction of some discrete mathematics at the school level? School mathematics all too easily degenerates into a succession of meaningless routines. And discrete mathematics has characteristics that make it vulnerable to such degeneration.

First of all, if problems in discrete mathematics are given in their natural setting, they tend to be rather difficult—even for those who feel they should know how to solve them. This is beautifully illustrated, if at a somewhat higher level, by the first four problems in Lovász's classic text (Lovász 1979, pp. 13–14):

1. In a shop there are k kinds of postcards. We want to send postcards to n friends. How many different ways can this be done? What happens if we want to send them different cards? What happens if we want to send two different cards to each of them (but different persons may get the same card)?

2. We have k distinct postcards and want to send them to all our n friends (a friend can get any number of postcards, including 0). How many ways can this be done? What happens if we want to send at least one card to each friend?

3. How many anagrams can be formed from the word CHARACTERIZATION? (An anagram is a word having the same letters, each occurring the same number of times; this second word does not have to have a meaning.)

4. (a) How many possibilities are there to distribute k forints [Hungarian coins] among n people so that each receives at least one?

   (b) Suppose we do not insist that each person receives something. What will be the number of distributions in this case?

The educational value of much simple discrete mathematics lies precisely in the fact that it forces students to *think* about very elementary things, such as systematic counting. However, this can easily be undermined by the fact that most mathematics teachers feel obliged to "help" students solve hard problems by reducing the solutions to a number of manageable and predictable steps, or rules, *requiring an absolute minimum of thinking*. If one adopts a curriculum that is too demanding, or if one fails to develop learning paradigms to guide the way the material is taught (cf. Freudenthal 1978, pp. 192–201), it will soon degenerate into a collection of predictable problems in stereotyped settings designed merely to trigger some relevant routine. Its educational value will then be lost.

Second, though the subject matter of discrete mathematics is elementary in the sense that it has few technical prerequisites, much of it is intrinsically *unnatural* to the average adolescent, for example, the pigeonhole principle (cf. Freudenthal 1978, pp. 210–14); trees and weighted, or directed, graphs; the limitations of "greediness" when seeking optimal solutions. Much that *can* be made accessible requires very careful teaching if it is to become natural, for example, the product rule for counting (cf. Freudenthal 1978, pp. 201–10); the trick of counting a larger set in two ways to solve for a subordinate unknown; matrices; the difference between, and the relative importance of, recurrence relations and closed formulas; the difference between an apparently correct algorithm and a proven correct one.

Third, it is frequently asserted that discrete mathematics has the advantage of having many apparently accessible real-world applications. These are accessible only in the sense that the problem one is trying to solve can be understood (in some suitably simplified form) and that a solution can be effected using some simple *given* algorithm. The relevant algorithm can be *implemented* by high school students, though the mathematical ideas behind it may well be out of reach. Thus the fact that the algorithm does what it claims to do has to be taken on trust. Including this kind of application in the high school curriculum makes it look impressive and may give it a modern feel, but we must remember that such algorithms tend to become obsolete sooner rather than later. The algorithms themselves are therefore of limited interest. What matters most for the students' long-term mathematical education are the mathematical ideas behind the design, construction, and analysis of such algorithms and the experience of applying some of these ideas to devise and improve their own simple algorithms.

At the introductory level discrete mathematics should be based on a handful of important *ideas*, not on theories or standard algorithms. It should concentrate on problems that require students to think in contexts that are familiar and natural, and it should not resort to poorly understood routines. The problems studied should be chosen to bring out a small number of central techniques. "If you teach geometry you do not prove a random sample of theorems. You select those theorems from which you can dominate the subject" (Engel 1983, p. 313). Aye, and there's the rub! Identifying the appropriate central techniques can be tricky.

> Take the rise and fall of the flowchart. Once they were considered essential. Later they were called useful, still later useless, and today they are considered harmful. But it takes some effort to get rid of the flowchart paradigm if you were raised on it. (Engel 1983, p. 313)

This is presumably one reason why Lovász, in the first quotation of this chapter, expects that "it will take much more work" before we can hope to reach a consensus as to which central techniques are appropriate at the school level.

To sum up, if we wish to exploit the educational advantages of easily understood problems that force students to think, then we must produce a revised curriculum that does not try to do *too much*. We must also develop suitable learning paradigms to help teachers teach new topics successfully.

One can build only on those intuitions of number and space that have already been developed. Thus the content and the context of school mathematics is constrained by the students' existing mental universe. Preparatory work is therefore important. If one wants to explore the art of counting in high school, then students must already have had experience with lots of intuitively structured counting, for example, seeing the eight corners of a cube as "four on top and four on the bottom" and the twelve edges as "four around the top and bottom and another four joining them up." If one wants to introduce ideas related to graphs and networks, one should avoid abstract graphs drawn on a page and concentrate initially on familiar networks, such as those formed by the vertices and edges of familiar polyhedra or by schematic road maps. Counting the *faces* of a dodecahedron is fine ("one on top plus five round it and the same on the bottom"), but counting *edges* reliably probably requires some crude form of the product rule ("twelve faces, with five edges round each face, and each edge counted twice").

Finally, when choosing algorithms for general consumption, we must resist the temptation to saturate students with useful but poorly understood examples. All that is needed is *one* thoroughly accessible problem with a variety of instructively different solutions that among them exemplify a number of central principles. Ideally the problem should be such that students can begin by trying to construct a solution of their own, can experience firsthand some of the difficulties that arise—even for a computer—with simple methods as the size of the problem grows, can improve on these simple methods, and can eventually meet and appreciate an elementary, yet surprisingly effective, standard algorithm.

POSSIBLE CURRICULUM CHANGES

Roughly speaking, the topics I would advocate for all, or nearly all, students are the following:

- Those aspects of the art of counting that lead to the addition and multiplication rules
- The exploration of simple geometrical counting problems to cultivate accuracy, to bring out the advantages of working systematically, to practise recognising simple sequences, and to give an intuitive feeling for recurrence methods
- Some frequently neglected work with integers
- A few simple algorithms
- Enough work with polyhedra and schematic maps to convince students of the ubiquity of networks and to do applications of simple counting

As someone who lives and works outside the United States, I cannot be sure how much of each topic is appropriate. I shall nevertheless give a fuller list of topics that, on the basis of my own work with children and college students, I can envisage being taught successfully to some subsets of school students. Much of the material is in fact familiar and can be begun as part of their existing numerical and geometrical work. Nevertheless, the list may suggest fresh connections and new emphases.

Counting (grades 1–12)

The art of counting should form a continuous thread throughout each child's education. It includes sequential counting, careful systematic counting, grouping and using multiplication to count arrays or grouped collections—often in two ways, formulating and applying the addition and multiplication rules and the standard trick of double counting. At each stage there should be lots of combinatorial word problems with the emphasis on thinking and common sense rather than on standard methods of solution.

Integers (grades 6–12)

Work with integers includes factorization and numbers in factored form—stressing connections with structured counting. This leads later to using factorization trees as a method and, for some, to finding factor lattices a source of challenging problems; to recognising squares and triangular numbers especially in factored form (e.g., $10 = (4 \times 5)/2$); to using integers as a natural source of counting problems (How many two-digit numbers use a 5?), leading for some to sequences (How many n-digit numbers use a 5?); and to finding greatest common factors—first by factorization and later using the Euclidean algorithm.

Sequences (grades 6–12)

Sequences can be generated by iterating simple arithmetical rules or from simple counting problems with parameter n—initially in verbal rather than symbolic or algebraic form. One advantage of counting problems is that errors for small values of n tend to accumulate, thereby emphasizing the need for accuracy and a systematic approach. Students should learn to recognize simple linear and quadratic sequences; to distinguish between an iterative scheme, $(u_1, \ldots, u_{n-1}) \rightarrow u_n$, and a formula for u_n in terms of n; to be aware of similarities between Fibonacci numbers and powers of 2; and to have some idea about rates of growth—linear, polynomial, and exponential.

Geometry (grades 4–12)

Geometry provides a rich source of natural counting problems: grids, square and triangular dot lattices, points and lines in a plane, other planar figures, polygons, polyhedra, crossing points and regions, and so on. There is also scope for some work on schematic representation of information by weighted networks with simple scanning for the shortest paths.

Recurrence (grades 6–12)

Recurrence is not so much a separate topic as a recurring theme! Students have to learn that when counting—or sorting, or searching—it pays to reduce the solution of any problem to a simpler version of the same problem, to reflect on what one has done, and then to express this reduction process in the form of a recurrence relation or algorithm. (For example, each factor of $60 = 2^2 \times 3 \times 5$ is either divisible by 5 or it is not; so each factor of 60 is either 5 times some factor of $2^2 \times 3$ or is itself a factor of $2^2 \times 3$. Hence, $2^2 \times 3 \times 5$ has exactly twice as many factors as $2^2 \times 3$. Similarly, $2^2 \times 3$ has exactly twice as many factors as 2^2. Now 2^2 has 3 factors. So 60 must have $2 \times 2 \times 3 = 12$ factors. Similarly, $420 = 2^2 \times 3 \times 5 \times 7$ must have exactly $2 \times 2 \times 2 \times 3 = 24$ factors.)

Algorithms (grades 8–12)

One should not try to do too much; two or three pregnant examples, or paradigms, are all that is required. Finding greatest common factors is an excellent example of an algorithm because there is a simple solution—factor both numbers and pick out the greatest common factor. This is easy to do by hand for small numbers, but it is harder to program and slows up visibly as the numbers get larger. Yet there is also an unexpectedly simple, and amazingly effective, alternative—the Euclidean algorithm.

My own preference for a rich example in which to explore combinatorial algorithms would be one for sorting. The problem of sorting names or words or numbers into their natural order is intuitively clear, and the difficulties inherent in simple methods can be experienced firsthand—by sorting, say, first 10 and then 100 book titles into alphabetical order. Several simple and effective algorithms, as well as some subtler ones, then arise.

Probability (grades 7–12)

This is a thoroughly artificial world that often annoys genuine statisticians. But once we have taught students simple counting, this topic can be used to help them get a feeling for *theoretical* probability without being confused by treating *observed* relative frequency as if it were the same as hypothesized "probability." The work provides an opportunity both to apply counting skills and to exercise the manipulation of fractions.

CONCLUSION

Determining the role of computers in discrete mathematics at the school level is exceedingly delicate. Computers grind out *numerical* answers, whereas what one needs in teaching discrete mathematics are *structured* answers. Numbers generally tell the novice nothing until she or he begins to see *how* they arise. The act of using a machine to grind out answers seems to make it difficult for students to think about anything other than the numerical answer (or the lack of one). Mathematics educators have a responsibility to explain why this is inadequate. In the United Kingdom most mathematics teachers do not understand the distinction between *numerical* and *structured* answers, what this distinction has to do with, say, the relative significance of fractions and decimals or what it has to tell us about the importance of algebra, to say nothing about the fundamental distinction between *numerical* precision (approximation) and *mathematical* precision (exactness in principle) and the characteristic mathematical methodology that results from this distinction. All this suggests that tomorrow's curriculum must be specifically designed to′ emphasize the difference between *meaningless numbers* and *structured answers* that give some insight into how they arise.

Computers are here to stay. They must be used, but in a controlled fashion. The educational problems they present must be clearly understood, and any changes we make must be framed to guarantee that the *anti*mathematical potential of computers is minimized.

> *Practise yourself, for heaven's sake, in little*
> *things; and thence proceed to greater.*
>
> Epictetus

REFERENCES

Engel, Artur. "Algorithms." In *Proceedings of the Fourth International Congress on Mathematical Education*, edited by Marilyn Zweng et al. Boston: Birkhäuser, 1983.

Freudenthal, Hans. *Mathematics as an Educational Task*. Dordrecht, Holland: Reidel, 1973.

———. *Weeding and Sowing*. Dordrecht, Holland: Reidel, 1978.

Lovász, László. "Algorithmic Mathematics: An Old Aspect with a New Emphasis." In *Proceedings of the Sixth International Congress on Mathematical Education*, edited by Keith Hirst and Ann Hirst. Budapest: Janos Bolyai Mathematical Society, 1988.

———.*Combinatorial Problems and Exercises*. Amsterdam: North Holland, 1979.

Poincaré, Henri. *Science et méthode*. Paris: Ernest Flammarion, 1914.

3

Strengthening a K–8 Mathematics Program with Discrete Mathematics

Claire Zalewski Graham

THE NCTM *Curriculum and Evaluation Standards for School Mathematics* contains a standard on discrete mathematics for grades 9–12 (NCTM 1989). But is discrete mathematics appropriate for an elementary or middle school mathematics program? Definitely, if we are truly preparing our students for the technological and information-oriented twenty-first century. During a child's elementary years the foundation is laid for strands that will be studied in depth in the future. One such strand is discrete mathematics. Discrete mathematics is not a new branch of mathematics that must be added to the existing curriculum. Rather, it is a collection of topics that most elementary teachers know something about and almost certainly already teach. These topics include counting techniques, sets, logic and reasoning, patterning (iteration and recursion), algorithms, probability, and networks. Each of these topics will be described briefly emphasizing problem solving with manipulatives. The illustrations are meant to give teachers additional ideas to implement in the classroom. It is hoped that they will be a catalyst for further exploration and development.

COUNTING TECHNIQUES

Counting techniques are used to solve a variety of problems. At the elementary school level, the emphasis should be on problems that use manipulatives or diagrams. Active involvement by the learner is essential. Using a formula to solve a problem is not appropriate for these children. After exploring numerous activities at the concrete level, most students should be ready to learn to use, or be familiar with, tools such as tree diagrams, the fundamental counting principle, and the pigeonhole principle to solve problems. Developing these strategies at a formal level is appropriate only after a great deal of concrete exploration and investigation. Two examples of problems using a counting technique to determine a solution follow.

1. Using three different colored cubes, how many different ways can you arrange the cubes in a row?

   In the primary grades children can investigate this question using cubes. Through trial and error a child may demonstrate the six different arrangements. The children can use one-inch graph paper to record the different arrangements with felt-tip markers or crayons. In the intermediate grades students can use a tree diagram to help them visually account for all possibilities (fig. 3.1).

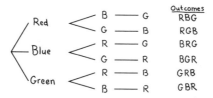

Fig. 3.1. Tree diagram

2. As a variation of this problem, plan the sandwiches for a picnic. You can use tuna fish or chicken salad on whole wheat or white bread, with any *one* of the following items: lettuce, tomatoes, onions, or pickles. How many different ways can you make the sandwiches? [*Answer*: 16 different sandwiches]

After students have had many opportunities to use concrete materials and tree diagrams to solve simple problems, a natural next step is the discovery of the fundamental counting principle, which can be used to solve similar but more complex problems. This principle states that if an event can occur in *a* ways and if, after it has occurred, a second event can occur in *b* ways, then the first event followed by the second event can occur in *a* times *b* ways. This principle can be extended to any number of events. Consider the following examples:

- If automobile license plates display six digits, what is the total number of different plates possible if only the digits 0 through 9 are used and repetitions are allowed? [*Answer*: 1 000 000]
- If automobile license plates display three letters followed by three digits, what is the total number of different plates possible if repetitions are not allowed? [*Answer*: $26 \times 25 \times 24 \times 10 \times 9 \times 8$]

The pigeonhole principle is another useful counting technique. One version of the principle can be stated, "If a set of pigeons is placed into pigeonholes and there are more pigeons than pigeonholes, then some pigeonhole must contain at least two pigeons." An example of this principle follows:

- Twelve red checkers and twelve black checkers have been placed in a bag. Without looking, how many checkers must you remove from the bag to be sure that you have five checkers of the same color?

To solve this problem, the students must consider the worst possible scenario: The first two checkers removed may be one of each color, and this may continue until four of each color have been removed. The ninth checker removed is sure to make five checkers of one of the colors.

Chapters 5, 7, and 8 contain more instructional ideas for developing counting techniques with children.

SETS

Set theory is another topic in the study of discrete mathematics. Currently, many primary-grade activities using manipulative materials explore aspects of set relationships and operations. Set concepts can be investigated throughout the elementary grades with formal vocabulary being introduced in the middle grades.

As early as kindergarten, children begin to form sets when they sort and classify objects. In simple terms, a set is a collection of objects that is so clearly described that it can be determined without question if another object belongs to that collection. Commercial materials, such as attribute blocks and Cuisenaire rods, or student- and teacher-collected materials, such as buttons, shoes, bottle caps, jar lids, pasta, keys, nuts and bolts, and spools, can be used for sorting and classifying.

Initially, objects are sorted by one attribute to form two disjoint sets. For example, a collection of buttons can be sorted by the attribute of color. Either the button is white or it is not white. The white buttons can be placed on a white mat, the nonwhite buttons can be placed on a colored mat. This illustrates two disjoint sets. Disjoint sets can also be used as an introduction to the concept of addition: If there are three white buttons and two green buttons, how many buttons are there altogether? A question like this suggests that addition can be interpreted as the combining or joining of two disjoint sets.

Next, the buttons can be sorted by two attributes—holes/no holes and white/not white. The concept of a loop or Venn diagram (using yarn or plastic hoops) can be introduced by showing that one button can be a member of both sets, thus creating a need to overlap the loops.

In the intermediate grades students can work with Venn diagrams with three intersecting circles and solve such problems as the following:

- Given the thirty-two-piece set of attribute blocks (four colors, four shapes, two sizes), place the pieces in the Venn diagram shown in figure 3.2. Will all the pieces in the set be placed inside the circles? Why or why not? If not, list the pieces that will not be included. Can the circles be labeled so that all thirty-two pieces could be included inside the three circles?

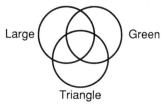

Fig. 3.2

This type of experience lays the groundwork for more involved problems to be investigated in the middle grades. For example, suppose a survey of 200 high school graduates yielded the following results:

80 students have studied French

89 have studied Spanish

64 have studied German

45 have studied French and Spanish

23 have studied French and German

35 have studied Spanish and German

3 have studied French, Spanish, and German

Questions to encourage the development of a Venn diagram might include these:

1. How many students have studied French but neither Spanish nor German? [*Answer*: 15 students]
2. How many students have studied French and German but not Spanish? [*Answer*: 20 students]
3. How many students have studied none of these three foreign languages? [*Answer*: 67 students]

LOGIC AND REASONING

The ability to reason logically is a skill necessary for daily living. Asking children to justify their thinking helps them clarify their reasoning. Useful questions are these: Why? How do you know? What makes you think about it that way? Is there another way? Questions that expect yes and no responses do not require students to explain. Low-level questions require students only to have knowledge and comprehension of the content. Encouraging students to analyze, synthesize, and evaluate the concepts will help them understand that mathematics makes sense and is not a collection of unrelated facts. Lessons in logic and reasoning can also be an integral part of every other content area in an elementary school curriculum.

Manipulative materials can be used effectively for activities requiring logic and reasoning. At the elementary school level, students can use manipulative materials and other models to help them justify their thinking. They provide concrete experiences for children to demonstrate what could otherwise be very abstract mathematics for students in the elementary grades.

All/Some/None

As young children sort and classify materials (wearing apparel, commercial objects, or "junk" collections), have them sort the objects in ways that allow discussion of the data using the words *all, some,* or *none.* Here are two examples:

1. Sort the sneakers worn by the children in the classroom. The result may lead to statements like these: All the white sneakers have laces. Some of the blue sneakers have laces. None of the black sneakers have Velcro.
2. Sort a set of cans. The results may yield these statements: All the cans have labels. Some of the cans contained food for pets. None of the cans have lids.

Constructing a class graph with information collected from the students presents data from which to draw logical conclusions. Depending on grade level, ask students either to discuss their findings or to write them first and then discuss them. Ideas for graphing include birth months, household pets, hair color, eye color, favorite subject in school, and favorite sport. The concepts of *all, some,* and *none* are appropriate in the discussions of these graphs.

If . . . Then

A concrete way to introduce a missing addend in addition problems is to use chips or other counters. Place a quantity of chips (e.g., five chips) on a mat. Ask the student to observe what is on the mat. Cover some of the chips. Ask, "If you now see three chips on the mat, then how many chips have been covered? How do you know that?" Children can work with a partner to gain additional experience with this activity.

Sequencing

Cut out from the newspaper popular comic strips that children can understand. Cut the frames apart and place them in an envelope with the title of the comic strip on the outside. Ask students to place the frames in a logical sequence and justify their responses. Students may want to draw their own picture sequences, cut them up, and challenge others to put them in order.

Cut apart short stories, created by either the teacher or the students, and give them to the children to put back together in the correct sequence. The sequenced story can then be read aloud to the class by the students.

Logical Deduction

With concrete materials such as attribute blocks, students can play a game called "Guess My Block." Give clues that systematically eliminate blocks until one block is found that has all the given characteristics. Depending on the grade level, blocks may or may not be sorted as each successive clue is given.

Here are two examples of sets of clues using the thirty-two-piece set of attribute blocks.

1. It is large.

   It has four sides.

   It is not a square.

   It is blue.

   It is _____. [*Answer:* Large blue diamond]

2. It is not blue or it is not small.

   It is a circle or triangle or square.

   It is not large.

   It is green or red.

   It is not a circle.

   It is not a four-sided shape.

   It is red.

   It is _____. [*Answer:* Small red triangle]

Children enjoy writing clues for other students to solve.

Inductive Thinking

Inductive thinking can begin as early as the primary grades when odd and even numbers are introduced. Manipulatives such as cubes or chips can be used to demonstrate concretely which numbers are even and which are odd. Shading graph paper can also be used as a semiconcrete model. Once a pattern has been established, students can be encouraged to extend their thinking so they can predict which large numbers will be even or odd.

Again, using the same concrete or semiconcrete materials, addition and multiplication of odd and even numbers can be explored. Once the students have reached a hypothesis, they may want to use a calculator to test it with larger numbers. Problems with more than two addends or factors can also be explored.

PATTERNING

Iteration is a technique that can be used to generate a pattern. Iteration means repeating a procedure over and over to develop a sequence.

Starting at the concrete level, attribute blocks can be used to generate a sequence. The simplest sequence is a one-difference train. Children work in groups of four for this activity. The first person selects any one of the blocks and places it on the table. The second person selects a second block that has only one attribute different from the first block and places it next to the first block. The third person selects a third block that is one attribute different from the second block and places it next to the second block. This train can be continued until either all blocks have been used or it is impossible to place another block. Additional challenges are to use all the attribute blocks in a circular train so that the last block used is also one different from the first block; to make a two-difference train using as many blocks as possible or all the blocks; to make a three-difference train with the same criteria as the two-difference train.

Numerical patterns using iteration are found in most elementary mathematics programs. Depending on the grade level, students may be asked to continue patterns such as these:

- 2, 4, 6, 8, _____, _____, _____
- 4, 7, 10, 13, _____, _____, _____
- 4, 8, 16, 32, _____, _____, _____
- 5, 5.5, 6, 6.5, _____, _____, _____
- 22, 22½, 23, 23½, 24, _____, _____, _____

Some examples for students in the intermediate or middle grades:

- Find the 50th or 100th term of a sequence.
- Given a sequence, determine what term a specific number will be; for example, given 5, 10, 15, 20, 25, . . ., what term of the sequence is 150?
- Given the first and seventh terms and assuming a constant difference, determine the first ten terms of the sequence; for example, complete this sequence: 21, _____, _____, _____, _____, _____, 63, _____, _____, _____.

Recursion is another technique used to solve problems when trying to describe future results by looking at previous step(s). The Fibonacci sequence, 1, 1, 2, 3, 5, 8, 13, . . ., is a familiar example of a recursive pattern. Each new term in the sequence is determined by the sum of the previous two terms. Students can be encouraged to look for the Fibonacci sequence of numbers in many fields, including science and music. Similar patterns can be created by starting with different numbers, 3, 3, 6, 9, 15, 24, ... The building of Pascal's triangle is another activity involving recursion that has much potential for the middle grades. Many patterns can be explored, including the generation of the numbers as they emerge in the triangle, the symmetry of the triangle, and the various operations on the numbers. Researching the many applications of this array of numbers can provide an interesting investigation for students to share with their classmates. For example, how does the arrangement of the numbers relate to the number of possible outcomes in a coin-tossing experiment?

Many elementary school students have had the opportunity to learn to write Logo programs. Logo recursion is a clever way to create designs and shapes by using procedures that call a copy of themselves as a step. The procedure SQUARE, illustrated in figure 3.3, uses recursion. Students should notice the need for another statement to control the turtle and make it stop.

TO SQUARE
FD 100 RT 90
SQUARE
END

Fig. 3.3. Logo square

Further exploration can create many interesting designs from basic polygons and the use of recursion.

ALGORITHMS

The investigation of algorithms in the study of mathematics can help students organize and structure their thinking. In essence, an algorithm is a sequence of instructions that, if followed for an operation (whether mathematical or not), will always lead to a result.

At the elementary school level, children can invent algorithms to solve addition, subtraction, multiplication, and division problems. In the middle grades, students can investigate the Euclidean algorithm as a way to find the greatest common factor of a set of numbers.

In addition to these numerical algorithms, many nonnumerical algorithms can be considered and explored. Children can sequence a given set of pictures. For example, the pictures in figure 3.4 illustrate in random order the steps to follow when feeding a pet cat.

Fig. 3.4

Depending on grade level, students can be asked to illustrate a sequence, write a list of steps in a sequence, or draw a flowchart to show a sequence. Some situations to explore that require algorithmic thinking are brushing teeth, making a peanut butter and jelly sandwich, buying a birthday card for a relative, walking a route from home to a friend's home, wrapping a gift, writing a letter, dialing a telephone, and changing a flat tire.

Since a systematic approach to task accomplishment is essential in the real world, teachers are encouraged to start this process with children as early as possible.

PROBABILITY

Will you need to wear a raincoat tomorrow? What are the chances that your favorite baseball player will hit a home run this week? Probability has many applications in business, sports, and the sciences. Stockbrokers, airline staff, and weather forecasters make predictions in their work. The study of random happenings can start in the elementary grades, since many investigations and experiments can be carried out with manipulative materials. Children enjoy making predictions and having experiences with elements of chance.

At the early stages, children can experiment with materials to determine *experimental* or *empirical* probability. By actually conducting an experiment several times, children can determine the number of ways an event occurred compared to the total number of possible events. Children in middle school can still conduct probability experiments but can also begin to discuss the *theoretical* probability of an event.

Some experiments or activities to use as a basis to compute probabilities follow:

- *Dice tossing*

  Have children work as partners and toss pairs of dice 25 times. One student rolls the dice and the other records the sum of the dice for each roll on a tally sheet. A bar graph is drawn to show the results. Use discussion questions like these: What sum appeared the most? The least? Can you predict the results for 100 tosses? 300 tosses? All the partnerships then pool their data. How accurate were the predictions? Can you explain why this happens? (*Hint:* Look at the number of ways each sum can be obtained when tossing two dice.)

- *Weather predictions*

  Have students watch the weather report on television or read it from a daily newspaper to record the probability of precipitation (rain or snow) predicted each day. Discuss what probability means and observe what happens on days when the predicted probability is very high or very low. The class may want to chart the predictions and actual outcomes. After an extended period of weather watching, discuss the accuracy of weather predictions.

- *Coins and spinners*

  Carry out similar activities and experiments with coins and homemade or commercial spinners.

- *Playing cards*

  A variety of probability problems using playing cards can be acted out and discussed with intermediate or middle grade children. For example, place a standard deck of playing cards (no jokers) facedown on a table. One card is drawn. What is the probability that it will be a picture card?

NETWORKS

Networks or graphs—that is, figures that consist of points called vertices, which are connected by edges—provide excellent discovery lessons for students in the elementary or middle grades. Activities at this level should be kept informal so that the emphasis can be on developing the students' abilities to represent, explore, and solve problems using graphs. Analyzing explorations and verbalizing their thoughts will be useful in evaluating their reasoning. Later, students will explore applying graphs to a variety of disciplines and careers.

Transportation networks are a rich source of situations that lend themselves to graphical analysis. A road inspector is responsible for checking all roads within a town. The inspector needs to travel each road but would like to do it in the most efficient way so that she does not have to travel any road twice during her inspection tour. The inspector may pass through any intersection of roads as many times as necessary. Is it possible for a road inspector to do her job for the street networks in figure 3.5? Remember—no road may be traveled more than once. [*Answer:* All street networks in figure 3.5 can be traveled except *b*, *d*, and *f*.]

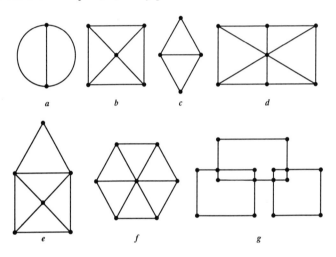

Fig. 3.5. Street networks

Organizing the data by recording them in a table may help students conjecture whether or not a network is traceable. Students may enjoy designing their own networks and having others decide if they are traceable or not.

A game called "sprouts," invented at Cambridge University in England in 1967 by John Conway and Michael Paterson, is based on graphs. The game is for two players and is started with a certain number of dots (two, three, or more). A move consists of connecting two vertices with an edge and placing a new dot anywhere on the new edge just drawn. There are two basic rules: (1) Each vertex can be the endpoint for no more than three edges, and (2) An edge cannot cross another edge. The last player able to draw an edge is the winner. Continue the game started in figure 3.6. How many more moves are possible?

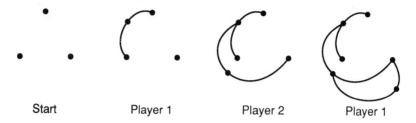

Fig. 3.6. Sprouts game

Since it can be observed that a graph divides a plane into regions, a natural follow-up to the study of graphs is the exploration of map coloring. The street networks in figure 3.5 can be the beginning of the exploration. The basic rule is to find a way to color the regions so that regions that share a boundary (a point is not considered a boundary) are not the same color. The goal is to shade each map with the fewest number of colors. How many colors are needed to color a map of the continental United States?

Networks offer interesting and challenging investigations. The experiences are rich and furnish opportunities for exploring, making conjectures, summarizing, and analyzing data. Additional examples are in chapter 4.

SUMMARY

Many concepts, problems, activities, explorations, and experiments from discrete mathematics are already present in elementary and middle school mathematics programs. What is needed is more focused attention on these topics. The examples cited here are a mere sample of the myriad of ideas that are in, or can be incorporated into, the mathematics curriculum. In time, additional discrete mathematics that is appropriate for elementary and middle school students can be added to the program.

The study of mathematics should be a broad and challenging experience for our students so that they will be prepared for the next century. Teachers are encouraged to introduce their students to the exploration and investigation of discrete mathematics topics to broaden their perspective of the study and application of mathematics.

REFERENCE

National Council of Teachers of Mathematics. *Curriculum and Evaluation Standards for School Mathematics*. Reston, Va.: The Council, 1989.

4

Graph Chasing across the Curriculum: Paths, Circuits, and Applications

Steven C. Althoen
Johanna L. Brown
Robert J. Bumcrot

D UE to the increasing need to apply mathematical concepts to real-world situations, the construction of mathematical models has become more important. One area of mathematics that is particularly valuable for creating models is the study of graphs. Many people have used concepts from graph theory without having studied them formally. We believe that students and teachers alike can enjoy properly designed activities that augment the student's knowledge of graphs and that this knowledge will supply the student with a foundation that will later enhance formal study of fields such as computer science and business.

It is worthwhile to study graphs at every level of the curriculum because of the following reasons:

- Graphs are *everywhere;* even young children encounter them when they do connect-the-dots drawings or play games like fox and geese.
- Graphs are *helpful;* they occur in road maps, constellations, circuit diagrams, and organizational charts.
- Graphs are *important;* they underlie many of the computer programs that make possible modern communications and information storage, processing, and retrieval.
- Graphs are *mathematical;* there are high-level mathematics research journals devoted entirely to graphs. Hundreds of articles appear each year.
- Finally, as we hope to show, graphs are *fun*.

A fundamental activity with graphs is the tracing of *paths*. This activity is especially appealing, since it can be done by very young children as well as by researchers developing answers to current problems. This chapter contains a number of activities to be used at various grade levels. Occasionally the activity presented for a particular grade level will depend on knowledge from a preceding activity. This will be noted. Sets of definitions are given that relate to the concept presented, and suggestions are offered for developing additional activities and for resources to expand and deepen activities for highly interested or motivated groups of students.

Definitions

A *graph* is composed of *vertices* and *edges*. Each *vertex* is represented on a drawing by a dot, and each *edge* is represented by a segment or arc connecting two vertices. Figure 4.1 shows a graph with nine vertices indicated by the dots labeled *A, B, C, D, E, F, G, H,* and *I.* It has fifteen edges connecting various pairs of vertices.

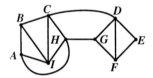

Fig. 4.1

Vertices that are connected by an edge are said to be *adjacent*. Thus vertices *C* and *D* are adjacent, vertices *A* and *H* are also adjacent, and so on. An edge is *incident* with the vertices it connects. Thus the edge from vertex *B* to vertex *I* is incident with vertices *B* and *I*. A *path* is a sequence of adjacent vertices. For example, in the graph of figure 4.1, there is a path from *A* to *B* to *C* to *H* to *G* to *D* to *C* to *I*. This path is denoted *A-B-C-H-G-D-C-I*. The path from *A* to *B* to *C* to *H* to *G* to *D* to *C* to *I* and then back to *A* begins and ends with the same vertex. This type of path is called a *circuit*. Two vertices are *connected* if there is a path between them. In figure 4.1 every pair of vertices is connected. An analysis of which vertices are connected describes the *connectivity* of the graph. The *degree of a vertex* is the number of edges that are incident with it. In figure 4.1, vertex *A* has degree three, *I* has degree four, and *E* has degree two.

ACTIVITIES FOR GRADES K–3

Kindergarten

Mazes and Mice. In this activity students follow a path without using drawing or other fine motor skills.

Draw a simple maze (fig. 4.2) with an entrance at one end and an exit with a goal at the other end. Add "hazards" (e.g., animals, water, other dangers) if desired for stimulating interest. Explain that the mouse would love to eat the cheese at the other end of the maze and that there is at least one correct path to that goal. Let each child guide an appropriate toy through the maze in an attempt to obtain the goal.

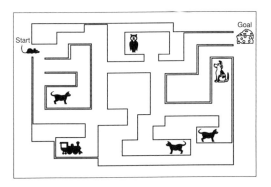

Fig. 4.2

Extension Activity. Create a large maze without creatures and goals on a shower curtain. Have students place their own hazards and rewards on the maze, and then guide a toy through the maze to its intended goal. Books and magazines that contain sample mazes are listed in the Bibliography.

Extension Activity. On the playground or other large, open space, have the students create a large maze by tying ribbons to chairs or wooden stands. They should place the usual hazards in the cul-de-sacs and challenge class-mates to walk through the maze to the goal. Velcro fasteners could be used to speed up the process if tying knots is a challenge. To create some excite-ment, a maze could actually be painted on the surface of the playground, or someone could photograph a human maze from the air (if it's possible to get up high enough).

First Grade

Connectit. In this activity students explore graph connectivity, compo-nents, and path search.

Have a child hold a Frisbee, or a large, round ring, in each hand. Ask a second student to take hold of one of these two Frisbees in one hand and a third Frisbee in the other. Let additional students join by taking hold of a Frisbee already in the game with one hand and a new Frisbee or one in use in the other hand. The goal is to have each student connected by Frisbees (vertices) to at least one other student. See figure 4.3.

While the first group is starting to connect on one side of the room, have another group of students begin the same thing on the other side of the room. Sometimes the two groups will join together and sometimes each will close up. If you have one large, connected group of students, ask students

Fig. 4.3

one at a time to let go of one Frisbee and take hold of another. Proceed until the graph disconnects. Then let a student not in the graph give the directions for when to let go and when to take hold.

Vary the game by adjusting the way students can combine at each Frisbee. Try the following variations:

- Each Frisbee must be held by at least two students.
- Each Frisbee must be held by exactly two students.
- Each Frisbee must be held by exactly three students.
- Each Frisbee must be held by at least three students.

As you try these variations, are there times when completing one or more of the variations is impossible? *(Remember the theorem that in any graph, the number of vertices of odd degree must be even.)*

When students are holding onto the Frisbees in differing numbers, they are representing models of differing *degrees* at each vertex. The degree of a Frisbee is the number of hands holding it. If exactly two students hold the same Frisbee, then the vertex is of degree two; if exactly three students hold the same Frisbee, then the vertex is of degree three. If one child holds a Frisbee with both hands and no one else holds the Frisbee, then that vertex is of degree two. (In graph theory an edge that joins a vertex to itself is called a *loop*.)

Extension Activity. Have some students who are not participating attempt to draw a picture of the graph formed by the participating students. Remember that each student is an edge and each Frisbee is a vertex.

Second Grade

Hookup. In this activity students extend the Connectit activity to include finding paths.

Have the students play Connectit as described above. When students have formed a nearly connected graph, have one student raise one Frisbee. This action raises the arm of each student holding onto that Frisbee. State that you are trying to create a "hookup" from child A to child D. The path could be from child A to child B to child C to child D. The activity begins when the first child A raises the Frisbee and says his or her own name and then says the name of student B. Each subsequent student in the path will raise the next Frisbee and say the name of the next student in the path. The student named must raise the Frisbee that is not held in common with the previous student and determine which way the path will proceed by calling out the next student's name. When child D is finally named, D will shout, "Hookup!"

Extension Activity. Name two starting students, A and G, for example, and two ending students, D and J. The students must race to complete the path from A to D and from G to J. The first hookup is the winner. Alternatively the winner might be the group with the most "direct" route from its starting to its ending student, determined by counting the number of edges (students) in the path.

Extension Activity. Have the students create the path to music, which may stop at any time during the development of the path. If the path is not complete when the music stops—and thus there is no hookup from student A to student D—the teacher names a new target student, and the path must proceed toward that vertex until the music stops again. Some graphs will be formed in which students A and D cannot be hooked up. You may wish to ask students to observe anything special about this situation and perhaps draw a model of what happened.

Third Grade

Round Trips. In this activity students explore circuits and graph diagrams. (Students should have played Connectit before they attempt Round Trips.)

Have the students play Hookup. Allow them to hook up with as many as four students at each vertex. (It may be necessary to use larger rings or rope rings for the vertices.) The goal of this new activity is not just to create a path from A to D but to also create a path from D back to A without repeating any of the vertices. A circuit without repeated vertices is called a *vertex-simple* circuit. When A is reached again, that student shouts, "Round trip!" This activity is more challenging when at least five students participate in the actual circuit.

Extension Activity. After the students have started with A and completed a round trip back to A, challenge them to complete a path in the form A-

B-C-D-B. When *B* is reached for the second time, that student shouts, "Round trip!" and the game ends.

The teacher should arrange some graphs where these activities are impossible (for example, see trees below), if students do not create such graphs for themselves.

Extension Activity. Complete a round trip using as many students as possible. At some point, the tangle of students and rings will become cumbersome, and pictures of the graph can be more easily manipulated on the chalkboard. Number the rings (vertices), identify the students (edges) holding them by letters or names, and transfer the graph to the chalkboard, using the following method: Draw a small numbered circle for each ring (it might be helpful to actually tag the ring with its number). Call on the students one by one using their assigned letter or name to draw the edge the student represents and the next ring. Ask the students to print their names along the edges. Each student named will let go of the two held vertices; when the last child has drawn his or her edge, the picture of the graph is complete.

Extension Activity. Write a circuit on the chalkboard, for example, 3-14-7-9-3, which can also be written as (3,14,7,9) without the repeated 3 in its name. Ask the class to write this circuit in other ways, leading them to, for instance, (7,9,3,14) and (7,14,3,9).

ACTIVITIES FOR GRADES 4–6

Definitions

A tree is a connected graph with no simple circuits. (Remember that a simple circuit starts and ends at the same vertex and has no vertex repeated.) This implies that a tree is a graph in which each two vertices are the ends of one and only one simple path.

Often a particular vertex is labeled the root and the tree diagram progresses up or down from this vertex. A tree with a root is called a *rooted tree*. The vertices at the end of the tree diagram are called *leaves*.

The following activities are presented in order of ease of understanding. Since students' knowledge at this level may vary, teachers may wish to complete earlier activities. For the Family Tree activity, students need to have some understanding of exponents and large numbers.

Fourth Grade

I think that I shall never see/A graph as lovely as a tree. In this activity students explore tree diagrams. The activities will be easier to understand if the trees are all drawn hanging down, as in figure 4.4.

Students in your class have probably been interested in "college bowl" quiz games or a basketball tournament or the baseball world series. It is interesting to determine how many different ways these tournaments could be played by considering the succession of winners and losers. Suppose *A* and *B* are in competition for the best three out of five games in a series. If *A* wins the first game, the tree diagram begins like the one in figure 4.4. Note that this diagram is just one of the two diagrams needed to represent the number of different ways the competition could be played: What if *B* won first? Complete the other half of the tree diagram.

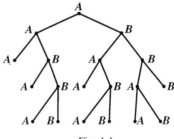

Fig. 4.4

Extension Activity. Change the rules so that the first team to win either two games in a row or a total of three games wins the series. Draw all the tree possibilities for this option.

Extension Activity. Draw a four-block-by-four-block grid of streets oriented north, south, east, and west, with a house on the NW corner and a school on the SW corner. See figure 4.5.

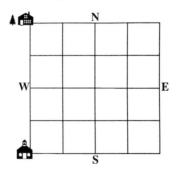

Fig. 4.5. Street grid

Starting at the house, decide at each corner which direction to go, following the rule that you may never return to a corner where you have been before. Using S for south, W for west, and so on, construct a tree of all possible routes to the school. For example, the shortest route is S-S-S-S.

Note that some walks can lead to traps from which you cannot escape under the no-revisit rule; for instance, S-E-E-N-W. You will probably want to modify this activity by making some streets one way or by inserting parks that eliminate several streets. In fact, the tree diagram for figure 4.5 is too large for anything but an extensive class project. You might want to find all paths that reach the school in a walk of four or six blocks. (Why not five blocks?)

Figure 4.6 is the entire tree for the apparently trivial two-by-two grid analogous to figure 4.5. For another variation, use an actual map of a small region around your school.

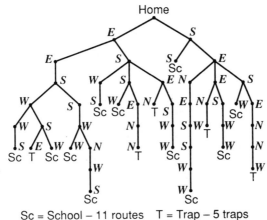

Sc = School – 11 routes   T = Trap – 5 traps

Fig. 4.6

During these constructions, introduce the terms *root* and *leaf*. The *depth* of a vertex is the number of edges from it to the root. In the example in figure 4.6, the depth of a vertex is the length of the walk to school by that route.

Fifth Grade

Untangling. In this activity students work on the classification of trees and discover that every tree with $n + 1$ vertices has exactly n edges. The teacher may want to prepare in advance a collection of trees, using large, multicolored beads or buttons for vertices and yarn for edges. The edges do not need to be long. These trees should be individually bagged to prevent them from becoming tangled with other trees. As teachers begin the discussion of trees with the class, they should take out one of these graphs, ask if anyone can lay out the graph as a tree, and follow with appropriate questions. Graphs could be laid out as in figure 4.7.

- How many different graphs are in figure 4.7?

Fig. 4.7

- Do the layouts look alike or different?
- How are they alike or different?

Answering the next questions is helped greatly by using beads of different colors at each vertex.

- Is each graph in figure 4.7 a tree?
- If the graph is a tree, can it be displayed as a rooted tree as shown in figure 4.4?

Students who have played the game Round Trip can be asked if there are any round trips in a tree graph. They should conclude that a tree is a graph with no round trips. Trace a simple path from one vertex to another vertex.

- Is there another path between these two vertices?

Discover that there does not seem to be an alternative path.

- Can figure 4.8 be laid out as a tree?

Figure 4.8, as it is shown here, has alternative paths between some of the vertices. This is because of the circuit in the figure.

- In figure 4.8, can you disconnect an end of one of the edges and put a vertex on the end of *that* edge to make the graph a tree?

Fig. 4.8

Extension Activity. Have the students use beads and yarn to make trees with 3, 4, 5, 6, and 7 vertices. Ask the students to count the number of edges each tree has. Collect this information in a chart, as in figure 4.9.

Does the number of edges seem to be consistent? If students have different numbers in the edges column, ask each student to draw the model of the graph and evaluate the graph. Is it really a tree? Students might think of edges as fencing sections and vertices as fence posts to help them understand. Try to think of other real-life examples.

Number of vertices	Number of edges
3	
4	
5	
6	
n	

Fig. 4.9

Extension Activity. Ask the students to make as many different trees as possible with 3, 4, 5, 6, and 7 vertices. Figure 4.10 shows some examples. Have them draw out all the variations and make a chart. Does there appear to be a pattern? No one has discovered a pattern yet.

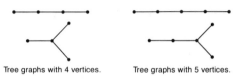

Tree graphs with 4 vertices.   Tree graphs with 5 vertices.

Fig. 4.10

● Are there other trees with 5 vertices besides those shown in figure 4.10?

Be sure to check for equivalent trees by moving the edges to see if the trees can be made to look alike. Ask the students to trim off an edge. How many edges and how many vertices are lost? Is this consistent with the idea that there is always one more vertex than edges?

Sixth Grade

Family Trees. In this activity students explore trees, measurement, and the idea of big numbers. It also relates mathematics to social studies.

Have students draw their family trees as far back as they can. Have the students guess how old their parents were when their children were born. Average these ages to determine the approximate number of years between generations. Discuss with the students that each person in the class has exactly two biological parents, and that each of those parents has exactly two biological parents. Make a chart of the number of people represented by each successive generation. One person has two parents, four grandparents, and so on. This generates the sequence 1, 2, 4, 8, 16, ... Students may notice, with a little help, that we double the number to get the next number in the sequence or even that the sequence is composed of successive powers of two. Ask questions about how many students there are in the class and how many parents they have. Then ask how many grandparents they have. How many great-grandparents? Eventually you may wish to ask the following questions:

- How many people could stand in this classroom? (Students may actually measure the room and calculate a reasonable number for this.)
- How many people could stand in the gymnasium or the cafeteria?
- How many people would it take to fill a city block?
- How many city blocks are there in this city?
- Could you go back enough generations of "parents" to actually fill the entire surface of the earth? Why didn't this actually happen?

Ask the students to make a tree-graph model of this situation. What factors allow us to conclude that a family "tree" is not an actual mathematical tree?

ACTIVITIES FOR GRADES 7–8

Seventh Grade

Constellations. Constellations of stars are usually presented as graphs. In this activity students use familiar constellations and change the location of the edges between the stars to form constellations with new names.

A graph of the constellation Leo is shown in figure 4.11. Leo has seventeen of its brightest stars as vertices. Have the students use the same seventeen vertices and make different graphs. When the graph forms a new shape, the student can give the constellation a new name.

Fig. 4.11

Extension Activity. Show the other constellations as graphs. Have the students repeat the activity above with different numbers of stars as vertices. Are there any constellations that actually have the same graph?

Spanning Trees of Maps. Make several tracings in pencil of a small section of a highway map. Designate some cities, towns, and crossroads as vertices and the sections of roads joining them as edges. Several edges may join the same two vertices in the original structure. Be sure to include enough roads so that the original structure is connected. Tell the students that a road section may be closed by erasing it, provided that the vertices that it joined are still connected by a path. The object of the activity is to close (erase) as many roads as possible. Students can proceed independently and discover that although they may find different answers, each correct answer is a tree with the same number of edges.

Eighth Grade

Huffman Codes. In this activity students apply tree graphs in a message-deciphering mode.

The teacher has intercepted a note in class. See figure 4.12. The tree tells the teacher that at least two students are familiar with Huffman codes. The teacher knows that the 0 means "left branch" and the 1 means "right branch." Armed with that information, the teacher begins to decipher the message. Starting at the top of the tree, the teacher follows the first numbers

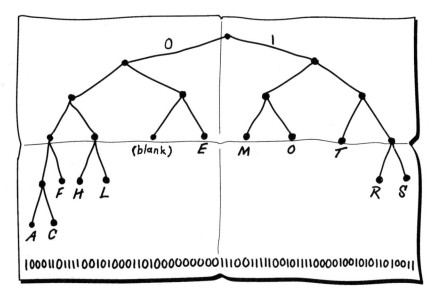

Fig. 4.12

in the message code. The 1, 0, 0 says to go right, then left, then left. Aha! The letter *M* is at this point. This is the first letter of the message. Returning to the top of the tree, the teacher follows the next part of the message. A 0, 1, 1, or left, right, right, command leads to the letter *E*. Another 0, 1, 1 leads to another *E*. Then a 1, 1, 0 leads to a *T*, and a 0, 1, 0 leads to a blank space. The first word of the message is *MEET*. Have the students finish decoding the message.

Extension Activity. After deciphering the message, the teacher is so taken with this new method of communication that he decides to pique the curiosity of the principal by sending a coded version of the message *I DESERVE A RAISE.* Help the teacher construct an appropriate tree for the letters used in this message. Tally the number of times each different letter is used in the message. The shortest paths should be used for the letters that appear most frequently. Since there are four *E*'s, the *E* should have a short code and should appear near the top of the tree. The tally is shown in figure 4.13.

Frequency	1	1	2	2	2	2	3	4
Letter	D	V	A	I	R	S	Blank	E

Fig. 4.13

Total the pair of smallest numbers, put their total above them, and connect them to their total. Continue to connect the two smallest numbers left until a tree is drawn that will be rooted at the top. Figure 4.14 shows the first three steps of this procedure.

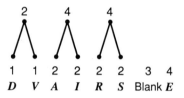

Fig. 4.14

This tree could be completed in many different ways. The four 2s could be combined in any order. The completed tree is shown in figure 4.15. When you have finished, you will have one number left. This number represents the total of the frequencies, or letters and spaces used, in this example, 17.

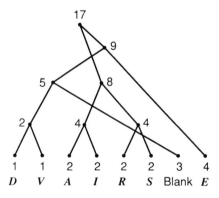

Fig. 4.15

Now redraw the tree so that it looks more like the earlier tree that the teacher intercepted in class. Write out the teacher's message for the principal. In case you need help, the message is encoded in figure 4.16.

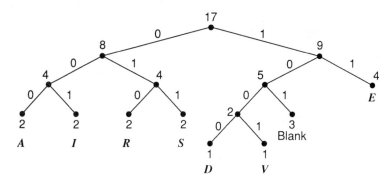

001101100011011110101001111010001010101000000101111

Fig. 4.16.

SUMMARY

In the spirit of the *Curriculum and Evaluation Standards for School Mathematics* (NCTM 1989), many activities based on graph concepts can be created for the enjoyment of young students. These experiences can be enhanced by using a variety of manipulatives and by presenting them in a game format.

BIBLIOGRAPHY

Fuller, Robert. *Amazement I*. Palo Alto, Calif.: Creative Publications, 1980.
_____. *Amazement II*. Palo Alto, Calif.: Creative Publications, 1980.
Koziakin, Vladimir. *Magic Maze Fun*. New York: Scholastic, 1985.
National Council of Teachers of Mathematics. *Curriculum and Evaluation Standards for School Mathematics*. Reston, Va.: The Council, 1989.
Stolow, Janet. *Fun Activity Book*. New York: Modern Publishing, 1983.

5

Primary Experiences in Learning What (As Well as How) to Count

Jane F. Schielack

A THREE-year-old girl, looking for someone to work with her on the computer, made her first foray into discrete mathematics. She began to list all the possible partners in her family. "Daddy could be my partner; or Mommy could be my partner; or my brother, Trey, could be my partner," she listed quickly. Then, after a moment of thought, without any outside prompting, she continued, "Or Daddy and Mommy could be partners, or Mommy and Trey could be partners, or Daddy and Trey could be partners." Satisfied that she had exhausted the possibilities, she proceeded to recruit her partner.

The *Curriculum and Evaluation Standards for School Mathematics* (NCTM 1989) calls for a K–4 mathematics curriculum that is broad in range of content, conceptually oriented, and developmentally appropriate with emphasis on students' active involvement in constructing mathematical ideas, applying mathematics, and developing the interrelationships of mathematical knowledge. Discrete mathematics, although not traditionally identified as a topic for the primary grades, can be investigated by young children in the context of familiar tasks and surroundings. Although they may practice counting given sets of objects, young children should also address, in particular situations, what *needs* to be counted.

Two of the major content strands suggested for kindergarten through grade 4 in the *Curriculum and Evaluation Standards for School Mathematics* are "Statistics and Probability" and "Patterns and Relationships." In grades 5–8, the *Standards* continues this content recommendation with a strand for patterns and functions and with separate strands for statistics and probability. For intermediate and upper elementary school students to be able to describe easily patterns of combinations and arrangements and to identify accurate sample spaces for determining probabilities, it would be helpful for them to have had many experiences in kindergarten through grade 2 in generating and recording arrangements within varying contexts. These

44

counting experiences can be based on actual situations familiar to children, as in the scenario above, and can be integrated into many of the activities that commonly occur in the primary-grade classroom.

COUNTING, WITH ORDER

Generating the Information

For an art project, ask kindergarten students to choose two pieces of construction paper, one for the picture and one for the frame, out of four possible colors. After the pictures are completed, display them on the floor or on the bulletin board. In an informal discussion, point out that Sue made a blue picture with a yellow frame and that John made a red picture with a white frame. Ask if the students think that these are all the ways that pictures and frames could be put together. Then help the students group the pictures by matching color arrangements. Some discussion may occur here about whether a blue frame with a yellow picture is different from a yellow frame with a blue picture. After the students decide that these *do* represent *different* arrangements, ask them if they think there are any other ways to put together a picture and frame.

At this point, if there is not already an example of a picture and frame of the same color, it may be suggested by a student or by the teacher. If the original directions were carefully worded so that phrases like "Choose two *different* colors for a picture and a frame" were not included, students will be open to this possibility. Students can then make examples of any picture and frame arrangements that they think are missing.

Summarizing the Activity

It is important for students to review their thinking processes after solving a problem and to share their procedures with one another. Appropriate process questions posed by the teacher in a large-group discussion can help students organize the information they have generated:

- *How did you decide that you had found all the color arrangements?* Answers to this question might represent a wide range of sophistication from "We just kept trying and couldn't make any new ones" to "We made as many pictures as we could with blue frames, then as many as we could with yellow frames, and so on." This last response leads directly to the next question.

- *How can we arrange our information to help us know that we have found all the possible color arrangements?* By grouping the pictures according to the color of the *frame,* the teacher can draw the students' attention to the fact that for each frame color, there should be either three or four picture colors, depending on whether the frame and picture are allowed to match. By organizing these results into four groups of three (if pictures and frames must be different colors) or four groups of four (if pictures and frames of the same color are allowed), the students can see the twelve or sixteen arrangements that can be generated with the four colors. The pictures could also be grouped according to the color of the *picture,* with each one having three, or four, possible frame colors. The teacher can then arrange the pictures in a 3×4, or 4×4, array, showing both ways of grouping simultaneously.

Although it is not appropriate with most primary students to discuss using multiplication for counting, this visual representation will provide information that they can recall in later activities asking them to relate multiplication to counting arrangements. They also will have experienced the importance of independent choices versus dependent choices. They must decide whether a piece of information, in this case a color, can be combined with itself or not.

Other Suggested Activities

Ask students to address similar questions about counting with order in a wide variety of settings. A group of three students putting on a puppet show or a play with three characters can make a chart showing how many different ways they could be assigned to the roles. Here it is important for students to realize that they are making dependent choices; they are restricted to using each person only once in each arrangement. They can then perform the plays using the six different arrangements.

Students can determine how many ways four sets of books or other materials can be arranged on the classroom shelves, placing one set on each shelf. For example, the dictionaries can be placed on the top shelf, the red readers on the second shelf, the paperback books on the third shelf, and the *Weekly Readers* on the bottom shelf. Or the dictionaries can be placed on the bottom shelf, and so on. Students should make some judgments about which of the twenty-four arrangements would be most desirable or practical.

Then the results from the first count can be revised in light of different restrictions. For example, the dictionaries, being the heaviest books, might need to be on the bottom shelf. How will that change the number of possible arrangements? With a visual representation of the possible ways to place the books on the shelves, students can see that only six of the twenty-four arrangements will satisfy the restriction and that the results will be the same as placing three sets of books on three shelves.

Other situations that call for counting with order include the number of ways a small group of students can be assigned to the class helper chart, the number of ways five students can be lined up to go to the library, and the number of ways the types of food—such as milk, main course, vegetables, and dessert—can be arranged in the cafeteria along the tray line.

COUNTING, WITHOUT ORDER

Generating the Information

In this activity, first-grade students, working in cooperative groups of three or four, are to determine how many different colors of paint they can make with pairs of four given colors: yellow, blue, red, and white. Before actually mixing the paint, each group predicts how many colors they think they will be able to generate. Using small plastic spoons, the students mix two spoonfuls of paint of each of two colors. It is important to specify clearly that equal amounts of paint be mixed in order to eliminate a wide variety of intensities of the same color. The colors may be mixed on a sheet of waxed paper, then transferred with fingers or brush to the record sheet. Depending on how much experience the students have had with organizing information, they might develop their own method of recording results or use the chart in figure 5.1.

Most students will begin this activity by choosing colors to mix at random. They become so interested in mixing colors that it may be helpful for the

COLORS WE CAN MAKE		
First Color mixed with Second Color		Resulting Color

Fig. 5.1

group members to have assigned roles: one person chooses the two colors to mix, one person mixes the colors, one person records the starting colors on the record sheet, and one person records the resulting color.

When each group is sure that all possible colors have been mixed and recorded, the results are posted with the results of the other groups. With the charts on display, the groups are then asked to share their strategies and conclusions.

Summarizing the Activity

- *Did you predict that there were more or fewer colors than you found?* The predictions in early counting experiences are more or less random, since most students do not have a strategy on which to base their predictions. If they have done the picture-frame activity described previously, they may make their predictions based on those results.

● *Did every group generate the same number of colors?* If the teacher has been very careful in the wording of the activity, avoiding phrases like "How many *new* colors" in order not to limit the students' explorations, it is possible that a group will include the four original colors as part of their results, for example, two spoonfuls of blue and two spoonfuls of blue. This type of exploration should be encouraged, since the increased number of results is an early experience in the difference between counting with independent choices and counting with dependent choices. Students can then engage in a profitable discussion of what the original question meant and if they think using the same color twice should be counted. The results might be grouped into new colors and original colors, and the answer to the question be presented as six new colors plus four original colors.

● *How did your group decide that you had found all the colors?* Answers to this question, as in the activity for counting with order, could range from "We just kept trying until we felt like that was all we could make" to "We mixed everything we could with red, then everything with blue, then everything with yellow, then everything with white." Using this response, the teacher might help the students list the results of following this procedure and compare it with how they recorded their final results.

● *Some groups have recorded red mixed with blue, whereas other groups have recorded blue mixed with red; why aren't both of these listed on each record sheet?* It is highly likely that these different orders will be represented on the record sheets. It should be very obvious to the students looking at the displays that in this instance *paint A mixed with paint B* is equivalent to *paint B mixed with paint A*.

● *How is what we are counting here the same as or different from counting the ways to put the pictures and frames together?* The goal in this discussion is to generate student responses that indicate an understanding of whether different orders of the same two colors are distinct or equivalent—whether the choices blue/yellow and yellow/blue are counted separately, as with the pictures and frames, or considered the same, as with the paint mixtures. This question works well as an informal assessment item, since the students' responses indicate whether they are dealing with each counting situation as an independent experience or whether they are able to find some relationships between the situations. The relationships are more easily seen by the students if the same four colors have been used in the two activities.

Other Suggested Activities

Given four ingredients such as chocolate chips, raisins, peanuts, and shredded coconut, how many different trail mixes can be made with exactly two ingredients? With exactly three ingredients? Ask the students to measure and mix the ingredients, organize the results, display the mixes in clear plastic bags on the bulletin board, and later sample them. Other situations that call for counting without order include selecting vegetables to make vegetable soup, selecting a small group of students from a larger group to sing a song, and choosing some flowers from a collection of different kinds of flowers to make a May basket.

CONCLUSION

Each of the counting activities should begin with predictions of possible results and be followed by questioning and student discussion that focus on the procedures used to find the combinations, the procedures used to decide when all the combinations have been found, and the similarities or differences between each activity and other counting situations that have been experienced.

The benefit of engaging primary students in discussions promoted by these activities is the development of their awareness of *what,* as well as *how,* to count. Certain attributes of counting situations—such as whether different orders are counted separately or not and whether choices are dependent or independent—that historically have been stumbling blocks for older students can be experienced by younger students in settings that will give them visual, auditory, and tactile references. Early experience in the need to identify such characteristics when counting provides a meaningful basis on which to build the symbolic representations of counting procedures.

REFERENCE

National Council of Teachers of Mathematics. *Curriculum and Evaluation Standards for School Mathematics.* Reston, Va.: The Council, 1989.

6

Discrete Mathematics in the Traditional Middle School Curriculum

James R. Hersberger
William G. Frederick
Marc J. Lipman

AS MATHEMATICS educators and mathematicians call for more discrete mathematics to be included in the school mathematics curriculum (NCTM 1989, Ralston 1985, Sandefur 1985), we must do more than suggest that new content areas such as graph theory, dynamic programming, and difference equations become part of the curriculum. Teachers must also use discrete methods whenever they are available to enhance learning in what is already a part of school mathematics. Consideration of discrete scenarios provides a natural method for improving problem-solving skills, practicing arithmetic, developing algebraic concepts, appreciation of algebra, and eliciting advanced mathematical ideas and formalism.

For example, a major obstacle in solving applied algebra problems is the inability to develop appropriate equations from given information (Clement 1982, Lochhead and Mestre 1988, Wollman 1983). Small-group problem-solving activities, focusing on discrete situations, can help greatly in developing equation-generating skills.

Consider the following problem (Lochhead and Mestre 1988, p. 132):

> I went to the store and bought the same number of books as records. Books cost two dollars each and records cost six dollars each. I spent 40 dollars altogether. How many books, and how many records did I buy?

When students are presented with simultaneous linear equation problems *before* they receive instruction in algebraic solution methods, they naturally employ a trial-and-error strategy. Although these solution efforts are seldom systematic, virtually all students can solve problems of the type just posed the first time they are confronted with one. It can be a lengthy process, but it is fruitful in the sense that the students solve a difficult problem *and* practice arithmetic in the context of doing something. For those who look for effective practice techniques, this benefit cannot be overemphasized.

As instructors, we encourage the development of heuristics that are more systematic; in fact, our students become quite proficient at solving problems using systematic trial-and-error procedures. However, we must take the eventual discussion further as we try to help the student develop satisfactory equation-generating skills.

For example, in the preceding problem, our students *know* that if B is the number of books and R is the number of records, then $B = R$. They have used that fact previously when using intuitive methods to solve the problem. They also *know* that $2B + 6R = 40$. They have used that fact in testing their guesses.

We do not wish to imply that solving a couple of problems will automatically prepare students to generate equations; on the contrary, it will still take a great deal of time and instructional effort to help students develop this skill. The point is that *explicit* discussion about students' intuitive use of appropriate equations in discrete settings allows students to develop strategies for determining appropriate formal equations.

There are benefits to teaching students who have already experienced similar situations in more inherently understandable contexts. Students' achievement and attitudes improve with greater understanding of the formal concepts, and the lengthy amount of class time spent initially on intuitive methods is more than offset by less time needed for instruction on the more formal techniques. For a more lengthy consideration of these ideas, see Butts (1985) and Demana and Leitzel (1988).

Discrete problems can also be powerful motivation for using and appreciating formal algebraic techniques. For example, use the following scenario, adapted from one presented by Thuente (1985):

> Four people wish to cross a river. They have a boat, but the boat holds at most two people. Each person takes a different amount of time to cross the river: person A takes 20 minutes, person B takes 15 minutes, person C takes 10 minutes, and person D takes 5 minutes. When two people are in the boat, it takes the maximum of their times to cross the river; that is, A and C together would take 20 minutes. What is the smallest crossing time required to get all four people across the river?

Place students in groups of three or four to work on this problem. When a group determines a solution, require each student to write a sentence describing the group's strategy for saving crossing time. The answer will virtually always be 55 minutes, and the strategy is to "send the fastest person every time, so you save time on the return crossing."

Then ask the same groups to repeat the problem using four people whose crossing times are 25, 20, 10, and 5 minutes respectively. After the expected response of 65 minutes, tell the group that these four people can make the journey in 60 minutes. The problem solvers must then develop a new strategy for saving time.

After a somewhat longer time than before, and often with more instructor intervention, the students decide that by sending the two quickest people first (C and D), sending one back (C), sending the two slowest across (A and B), sending the fastest back (D), and finally sending the two fastest again (C and D), the total crossing time with this sequence is 60 minutes.

Then assign one of the following tasks for homework: Middle school classes are asked to find crossing times (if possible) for which the first strategy is faster than the second. Algebra and prealgebra classes are asked to decide *when* each strategy is better given individual crossing times of $a > b > c > d$.

Students in the algebra and prealgebra classes generally discover the following information:

Strategy 1	Strategy 2
(A,D)	(C,D)
(D)	(C)
(B,D)	(A,B)
(D)	(D)
(C,D)	(C,D)

Strategy 1 can be seen to have a "cost" of $a + b + c + 2d$. Strategy 2 has a "cost" of $a + 3c + d$. The *difference* between the costs comes from comparing $(b + d)$ to $2c$. If the first quantity is larger, pick strategy 2. If the second quantity is larger, pick strategy 1. If they are equal, take your pick.

The river-crossing problem is a marvelous example of how algebraic formulation combined with the dynamics of small-group problem solving can provide a much greater understanding of a problem and its solution. Requiring students to write a sentence explaining the group's solution forces each student to be an active participant in the solution process.

In spite of its seemingly whimsical setting, this puzzle is a good example of the types of problems that attract professional mathematicians. This problem came from the problem corner of an operations research journal. Knowing that the problem is one that professional mathematicians have worked on gives the students a good feeling and reminds them of the power of their own strategies. In this instance, it also extends the problem, since the students do not yet know if they have the optimal solution.

CONCLUSION

The middle school mathematics curriculum is virtually barren of mathematical-concept development (Flanders 1987). A natural way to address the problem is to include more concept development in areas that will promote greater understanding of high school mathematics. Reasonable areas for focus include geometry, ratio and proportion, and algebraic situations in discrete contexts.

At their worst, discrete problems are either unsolvable until more sophisticated mathematics is encountered or simply unsolvable. However, even the more difficult or unsolvable situations are opportunities to "do" mathematics, and learning to deal appropriately with failure is also an important part of students' mathematical development.

At their best, discrete situations are solvable, provide problem-solving practice, and offer insight into situations or techniques that deserve further consideration. We believe that allowing students to use problem-solving strategies and techniques to understand problems *before* those problems are used to introduce efficient techniques or advanced concepts is crucial to the process of mathematical development. Returning to the same or similar problems at appropriate times in a student's mathematical career provides a richer context for understanding ideas, allows the instructor to introduce new classes of problems and mathematics in a familiar setting, and lets students operate like people who use mathematics in real life.

REFERENCES

Butts, Thomas. "In Praise of Trial and Error." *Mathematics Teacher* 78 (March 1985): 167–73.

Clement, John. "Algebra Word Problem Solutions: Thought Processes Underlying a Common Misconception." *Journal for Research in Mathematics Education* 13 (January 1982): 16–30.

Demana, Franklin, and Joan Leitzel. "Establishing Fundamental Concepts through Numerical Problem Solving." In *The Ideas of Algebra*, 1988 Yearbook of the National Council of Teachers of Mathematics, edited by Arthur F. Coxford, pp. 61–68. Reston, Va.: The Council, 1988.

Flanders, James R. "How Much of the Content in Mathematics Textbooks Is New?" *Arithmetic Teacher* 35 (September 1987): 18–23.

Lochhead, Jack, and José P. Mestre. "From Words to Algebra: Mending Misconceptions." In *The Ideas of Algebra*, 1988 Yearbook of the National Council of Teachers of Mathematics, edited by Arthur F. Coxford, pp. 127–35. Reston, Va.: The Council, 1988.

National Council of Teachers of Mathematics. *Curriculum and Evaluation Standards for School Mathematics*. Reston, Va.: The Council, 1989.

Ralston, Anthony. "The Really New College Mathematics and Its Impact on the High School Curriculum." In *The Secondary School Mathematics Curriculum*, 1985 Yearbook of the National Council of Teachers of Mathematics, edited by Christian R. Hirsch, pp. 29–42. Reston, Va.: The Council, 1985.

Sandefur, James T., Jr. "Discrete Mathematics: A Unified Approach." In *The Secondary School Mathematics Curriculum*, 1985 Yearbook of the National Coucil of Teachers of Mathematics, edited by Christian R. Hirsch, pp. 90–106. Reston, Va.: The Council, 1985.

Thuente, David J. "Brainteasers." *OR/MS Today* 12 (1985): 30.

Wollman, Warren. "Determining the Sources of Error in a Translation from Sentence to Equation." *Journal for Research in Mathematics Education* 14 (May 1983): 169–81.

7

The Pigeonhole Principle: A Counting Technique for the Middle Grades

Denise A. Spangler

T HE pigeonhole principle has been described by Dossey et al. (1987) as "a surprisingly simple existence statement that has many profound consequences." It is precisely this simplicity that makes it such an appropriate topic to introduce to students in the middle grades. The essence of the pigeonhole principle lies not so much in the mathematical computation needed to use it but in the mathematical thought processes that underlie its use. Encouraging children to think at this level of analysis helps them form a view of mathematics as a logical, interesting, thought-provoking discipline.

Since birthdays are of personal importance to children, the following example is a good starting problem:

> If I put all our names in a jar, how many names must I draw before I am certain that two of the people whose names I draw were born in the same month?

Most children know someone who shares their birth month, so they think that only two people would have to be chosen—themselves and the person who was born in the same month. (If a birthday chart is posted in the room, it might be helpful to remove it from sight.) Before proceeding, the students need to understand that we cannot control whose names are chosen. In other words, I want to know how many names I must pick so that no matter whose names I choose, I will be guaranteed that at least two of them were born in the same month.

Most children probably have never seen a pigeon coop or a rolltop desk and are unfamiliar with a pigeonhole, so it is helpful to modernize the concept for them in a concrete way. Since there are twelve months in a year, an egg carton makes a good model for this problem. Label the cups in the egg carton with the names of the twelve months. (An overhead transparency divided into twelve sections also works well). (See fig. 7.1.) Explain the problem in terms of the concrete model. When a person's name is drawn,

Fig. 7.1

a colored chip (or a penny) will be placed in the cup that corresponds to his or her birth month. The desired outcome is to have two chips in the same cup. I want to know how many names I have to draw before this *must* occur.

In light of the desired outcome, ask the children what is the worst thing that could happen. Children quickly catch on to the idea of playing "devil's advocate" in order to prevent the attainment of the goal. Suppose the first name drawn has a January birthday. The worst thing that could happen on the second draw is to get a month other than January, perhaps February. Using the egg carton and colored chips, place a chip in each section as the month is selected. When all twelve sections have chips (fig. 7.2), it becomes visually obvious that the thirteenth draw will result in two chips in the same month (fig. 7.3). The conclusion is that with an arbitrary number of people we would have to select thirteen names to be sure that two of the people would share the same birth month.

Fig. 7.2

Fig. 7.3

At this point it is instructive to employ the "look back" step suggested by Polya and review the problem with the students. The students should be encouraged to formulate a verbal explanation of the answer *thirteen* in terms of the original problem. Does this mean that if I drew your names out of a hat, I would have to draw *exactly* thirteen names before I got two people with the same birth month? Children should be able to see that this is not necessarily the case because the first name drawn may have an April birthday and so might the second. (This example can be made more meaningful by using the names of two students from the class who have the same birth month.) Other scenarios are possible as well. Ask the children to describe several other drawing sequences that would result in a match before the thirteenth draw. What, then, does our answer of thirteen draws mean? Students should recognize that on the thirteenth draw we are certain of getting a match, since even if the first twelve draws result in twelve different months, the thirteenth draw *must* be the same as one of the twelve months already drawn. Record this result as shown in table 7.1.

TABLE 7.1

Number of People with Same Birth Month	Number of Draws Required
2	13

Lead the students through another problem like the first, but this time tell them we want to find *three* people who share the same birth month. We can continue on from the results of our first problem, noting that after thirteen draws we have one chip in each of the twelve cups and one duplication. Here it is useful to have two colors of chips for visual emphasis. Suppose for the sake of convenience that the thirteenth draw was a January birthday. We again want to use the worst-possible-case approach to discuss what will happen on future draws. Following the same pattern as before, we find that after twenty-four draws we have two chips in each month. Thus it is the twenty-fifth draw that results in having three chips in the same month. Again, emphasize with the students that twenty-five draws are required to *guarantee* that we have three people with the same birth month, but it could happen sooner.

After entering these data into the table, divide the students into cooperative learning groups of no more than four students, give them an egg carton, some chips, and a calculator (for those who would like to do some conjecturing) and have them fill in the next few lines of the table. (See table 7.2.) Encourage the students to verbalize any patterns they see in the data.

TABLE 7.2

Number of People with Same Birth Month	Number of Draws Required
2	13
3	25
4	37
5	49
6	61

Students should state (in their own words) that to guarantee that there will be n people with the same birth month, we need to draw $12(n - 1) + 1$ names because $12(n - 1)$ draws will result in $n - 1$ chips in each cup, so one more draw will produce one cup with n chips in it. (Depending on the sophistication of the students, it may or may not be appropriate to introduce the algebraic notation for the general case.)

Once students are able to express this relationship verbally, check to see that they understand the concept by modifying the problem as follows:

> If I put all our names in a jar, how many names must I draw before I am certain that two of the people were born on the same day of the week?

Students should realize that since there are seven days in a week, the eighth draw would result in a duplication. What if I wanted to be sure I had three people who were born on the same day of the week? This would require fifteen draws because the first fourteen draws could result in two people on each day of the week. The fifteenth draw would have to be a repetition.

The following problems can be used as extensions and further examples of the pigeonhole principle. It is helpful to provide the students with some type of counting device and containers to help them keep track of their thinking. Small lima beans and condiment cups from restaurants make an inexpensive and versatile counting aid. A calculator is also recommended for those students who are immediately able to reach a solution or for those who want to verify that their result fits the generalization reached earlier.

1. A drawer contains unsorted black, brown, blue, and gray socks. If I select them in the dark, how many socks must I select in order to be certain of getting a matching pair? [*Answer:* 5 socks]

2. How many words must I choose from our spelling book to be sure that three begin with the same letter? [*Answer:* 53 words]

3. A bag contains pennies, nickels, dimes, and quarters. How many coins must I select to be sure I have four of the same type? [*Answer:* 13 coins]

4. If my compact-disk collection consists of eight albums by Kansas and five albums by Chicago, how many CDs must I select in order to get two by the **same** group? [*Answer:* 3 CDs]
   What if I want to have two CDs by **each** group? [*Answer:* 10 CDs]

5. Suppose there are 186 major-league baseball players hitting from .250 to .299. Why must at least four of them have the same batting average? [*Answer:* Since there are fifty batting averages from .250 to .299, to get four players with the same batting average, I would need at most 151 players. Therefore, there are at least four players of the 186 who share the same batting average.]

REFERENCE

Dossey, John A., Albert D. Otto, Lawrence E. Spence, and Charles Vanden Eynden. *Discrete Mathematics.* Glenview, Ill.: Scott, Foresman & Co., 1987.

8

Permutations and Combinations: A Problem-solving Approach for Middle School Students

Linda J. DeGuire

PERMUTATIONS? . . . Combinations? . . . Aaagh!" Many people have unpleasant memories of these topics. This chapter presents a problem-solving development of these topics that makes them enjoyable, understandable, and meaningful for both middle and high school students. The amount of time needed to present the material in an actual classroom will vary with the level of the students. Middle school students will need several class periods, whereas high school students may successfully assimilate the material in one class period. Prototypical solutions, teaching suggestions, and comments are interspersed throughout. In the problem solutions, the teaching suggestions appear in italics to distinguish them from the solution. The reader is strongly encouraged to read with pencil in hand—active participation through solving the problems is recommended in the exploration and development of the topics.

FUNDAMENTAL COUNTING PRINCIPLE

The fundamental counting principle asserts that if one task can be performed in m ways and a second task can be performed in n ways, then the number of ways of performing the two tasks is mn. This principle can be extended to any number of tasks. Most middle school students—and even many secondary school students—would not find such a formal statement meaningful or useful. However, if the principle is developed in relevant and concrete situations, many students will relate to it, as in the following example:

Problem 1. Barbara has 3 blouses and 2 skirts, all of which coordinate to make outfits. How many different outfits can Barbara make?

59

Solution. Call the blouses B1, B2, and B3, and call the skirts S1 and S2. *(With middle school students, use the words. Letters are used here to abbreviate the presentation.)* Then students can use the problem-solving strategy of listing all possibilities to arrive at the following list of outfits:

B1-S1	B2-S1	B3-S1
B1-S2	B2-S2	B3-S2

The list shows 3 groups of 2 outfits each. *(Regardless of the order the students use to suggest the combinations, write them on the chalkboard or overhead projector in an arrangement, as above, that suggests 3 groups of 2.)* Thus, Barbara can make 6 different outfits.

Alternative solution. Some students may find a graphic representation of the list more meaningful. In figure 8.1, each branch of the tree diagram represents one outfit. As with the list above, there are 3 sets of 2 branches each.

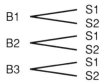

Fig. 8.1

Students need more examples of these types of problem situations, including those that extend the list or tree diagram to include three choices, as in this example:

Problem 2. A nearby restaurant offers five choices on its dinner menu— meatloaf, hamburger steak, chicken breast, pork chops, and shrimp. With each entree, the customer may choose one from each of the following two groups: french fries, rice, or baked potato; ice cream or Jell-O. How many different meals could be ordered at this restaurant?

Solution. Represent the entrees by M for meatloaf, H for hamburger steak, C for chicken breast, P for pork chops, and S for shrimp. Let the side dishes be F for french fries, R for rice, and B for baked potato and the desserts be I for ice cream and J for Jell-O. *(Give students an opportunity individually, or in groups, to list the different meals. Encourage them to first list partial meals of just entrees and side dishes. Point out orderly ways of making such lists. After the students have some of the combinations, suggest that they put dessert choices with each of the partial meals to make full meals. Finally, use the problem-solving strategy of listing all possibilities by pooling lists on the chalkboard or overhead projector. Arrange the meal combinations similar to the arrangement below to show 5 groups of 3 initially and then 15 groups of 2. The tree diagram in figure 8.2 shows the 5 groups of 3 and the 15 groups of 2 very clearly.)*

MFI	HFI	CFI	PFI	SFI
MFJ	HFJ	CFJ	PFJ	SFJ
MRI	HRI	CRI	PRI	SRI
MRJ	HRJ	CRJ	PRJ	SRJ
MBI	HBI	CBI	PBI	SBI
MBJ	HBJ	CBJ	PBJ	SBJ

There are 15 entree/side dish partial meals. Each of these 15 possibilities can be put with each of the 2 dessert choices, making 15 groups of 2. Therefore, 30 different meals are available at the restaurant.

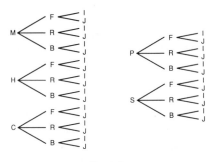

Fig. 8.2

Before more complex problems are introduced, it is important for the students to abstract the commonalities among these situations and find a more efficient procedure than listing possibilities or making tree diagrams. Many students discover the multiplicative nature of the process themselves. They readily see that the solution is reached by multiplying the number of choices at each position, or *slot*, together. Thus, in problem 1, 3 choices times 2 choices gives 6 outfits, and in problem 2, 5 choices times 3 choices times 2 choices gives 30 meals.

The fact that the product of the numbers of the choices for each slot yields the number of possible combinations is basic to the rest of the activities presented here. That is why it is called the *fundamental* counting principle. It is *very* important for the student to solve a variety of such problems by making lists or tree diagrams and counting the combinations. These solutions may be tedious, but they convince the student of the reasonableness of multiplying to find the number of combinations. When the students are sure that the results are true and do not need to continue verifying them with lists, problems like the following in which such a list would be formidable can be introduced:

Problem 3. A national organization wants to issue each of its members an ID code. The officers decide to use a four-character code that begins with a letter (not including O, in order to avoid confusion with the digit zero) and continues with three digits. They have 32 000 members in the organization. Will they be able to assign each member a different ID code?

Solution. There are 4 slots to fill here. The first slot (the letter) has 25 choices and each of the other slots has 10 choices (i.e., the digits 0 through 9). So the number of possible ID codes is $25 \times 10 \times 10 \times 10$, or 25 000. Since the organization has 32 000 members, they will not have enough distinct ID codes for each member. They need to redesign their plan and add another letter or another digit or change one of the digits to a letter. *(The question of whether the number of possibilities is enough to fit the condition or is more than is needed affords the opportunity to bring the problem solving closer to real life. Ask the students to modify the proposed plan—for example, the ID code—to be more realistic and to evaluate the proposed modifications. For example, in this problem the proposals of adding another letter or another digit will generate 625 000 and 250 000 different ID codes, respectively, each with five characters. The third proposal (changing one of the digits to a letter) will generate 62 500 ID codes, each with four characters. Unless the organization plans to double its membership in the near future, the last modification would require less computer memory space and less typing for each membership code. Thus, this code would be more cost-efficient than the other two proposals. However, if such an increase in membership is anticipated, reassigning and retyping codes and reprogramming computers for the extra character at a later time would be time-consuming and costly. In this situation, one of the other proposals should be considered. A variety of other parameters could be considered in the discussion.)*

PERMUTATIONS WITHOUT REPETITIONS

After students have learned and practiced the fundamental counting principle, they can tackle problem situations that involve permutations. In permutation situations, the order of the slots matters, as shown in problem 4. Initially, the repetition of elements should be avoided.

Problem 4. A class is having an election for class president. Four names are to be listed on the ballot—Jim, Karen, Linda, and Michael. How many ways can the names be listed?

Solution. Use the problem-solving strategy of listing all possibilities. *(Encourage students to make their lists in an orderly way. The order itself is not particularly important, as long as it gives all possibilities and makes sense to the student. Students usually need some practice with such listings, especially with small sets, such as the possible orderings of letters in the word* cat. *The practice is most useful after an introduction to permutations.)* Use J for Jim, K for Karen, L for Linda, and M for Michael. Make a list in four groups, one for each name. Within the group beginning with J, the other three names could be arranged in any of the following six ways: KLM, KML, LKM, LMK, MKL, MLK. The complete list of possible arrangements looks like figure 8.3. There are 24 possible ways (4 groups of 6 each) to arrange the four names on the ballot.

Fig. 8.3

Alternative solution. *(In many classes, the following solution may have already been proposed by some students. At this early stage of development, solving the problem in more than one way is essential. The less sophisticated solution procedure of listing all possibilities serves as a proof to convince the student that the more sophisticated procedure of multiplying the number of choices in each slot is believable and acceptable.)* First, use the problem-solving strategy of trying a simpler problem first. How many ways could the first two slots be filled? The first slot has four possibilities. The second slot has only three choices left. So, the first two slots could be filled in 4×3, or 12, different ways. How many ways could the 12 beginnings be finished? There are 2 names left. The third slot has two choices, and the fourth one has one choice. Thus, 2×1 ways exist to complete each of the 12 beginnings. Finally, putting these pieces together, we find that the names could be listed on the ballot in $4 \times 3 \times 2 \times 1$, or 24, ways.

The concept of the factorial of a number (e.g., $4! = 4 \times 3 \times 2 \times 1$) is frequently introduced as a shortcut for a product of a decreasing sequence of factors.

The introduction to permutations with a problem like problem 4 can be followed with a series of exercises to practice the newly learned procedure. The exercises can vary in situation and difficulty appropriate to the level of the students. Difficulty can be varied at least two ways—by increasing the size of the numbers involved (problem 5) and by including special considerations for certain slots (problem 6). *(Be careful about increasing the size of the numbers. The number of possibilities increases rapidly.)*

Problem 5. *(Increase the size of the numbers to the maximum possible on an 8-digit calculator without going into scientific notation or showing an* *error.)* How many ways can the letters of the word *workmanship* be arranged?

Solution. There are 11 slots to be filled, with 11 choices for the first slot and one less choice for each successive slot. Thus, $11 \times 10 \times 9 \times 8 \times 7 \times 6 \times 5 \times 4 \times 3 \times 2 \times 1$, or 11!, arrangements can be used, that is, 39 916 800 arrangements—a few more than it is reasonable to list.

Problem 6. *(Put extra conditions on one or more of the slots.)* How many different ways can the letters of the word *purchase* be arranged if each arrangement must begin with a consonant and end with a vowel?

Solution. There are 8 slots to be filled. Use the problem-solving strategy of breaking the problem into pieces—the first slot, the next 6 slots, and the last slot. The first slot has five choices, the consonants *p, r, c, h,* and *s.* The last slot has three choices, the vowels *u, a,* and *e.* For the middle 6 slots, 6 of the 8 letters are left from which to choose; the 6 letters can be arranged in $6 \times 5 \times 4 \times 3 \times 2 \times 1$, or 6!, or 720, ways. Thus, the letters of *purchase* can be arranged in $5 \times 6! \times 3$, or 10 800, ways, each beginning with a consonant and ending with a vowel.

COMBINATIONS

Though the order of elements is sometimes a consideration, it is irrelevant in many problems. For example, in Michigan's statewide lottery, the ticket purchaser tries to choose six numbers from the whole numbers 1–45 that will match the six numbers picked on Saturday evening. On Saturday, 45 numbered balls blow around in an air chamber until first one ball—then another—then another, and so on, falls down a chute. The chute is closed after six balls have fallen. Since only the six numbers are important and not their order, this is an example of a combination of 45 things taken 6 at a time. This example is too complex for most students to use successfully as an initial attempt to solve such problems. The technique of solving a simpler problem first should be useful here.

Problem 7. Suppose a special lottery requires choosing 2 letters to try to match the 2 letters that will be picked from 4 lettered balls A, B, C, and D. How many ways could a ticket purchaser choose the 2 letters?

Solution. There are 2 slots to be filled, with 4 choices for the first slot and 3 choices for the second slot. Using the fundamental counting principle, we see that there are 12 ways to choose the 2 letters. But wait. Does it matter whether A is chosen first and then B or B is chosen first and then A? No, it doesn't matter. This list of 12 ways has duplicates in it that differ only in the order in which the letters are arranged. List the 12 ways and cross out duplicates. Partitioning a set into subgroups of 2 suggests division.

The number of combinations of 4 letters taken 2 at a time is $(4 \times 3) \div 2$, or 6.

It is unlikely that students will be able to derive the method from this one example. They need further examples before being challenged to make a generalization.

Problem 8. Suppose a special lottery requires the ticket purchaser to choose 3 letters to try to match the 3 letters to be picked from 4 lettered balls A, B, C, and D. How many ways could the ticket purchaser choose the 3 letters?

Solution. There are 3 slots to be filled, with 4 choices for the first, 3 for the second, and 2 for the third. Thus, there are $4 \times 3 \times 2$, or 24, ways to choose the 3 letters. However, some of these ways are the same except for order. As in the solution to problem 5, list the 24 ways, and group together the ways that are the same except for order. (See fig. 8.4.) Each of the 24 ways falls into a subgroup of 6 that are alike except for order. Since partitioning a set into subgroups of 6 suggests division, the number of combinations of 4 things taken 3 at a time is $(4 \times 3 \times 2) \div 6$, or 4.

Fig. 8.4. Listing the 24 ways to arrange the 3 letters

Now challenge the students to see how they could have predicted the 6 subgroups in problem 8, and the 2 subgroups in problem 7. Being able to find this number without actually listing combinations that are the same except for order would shorten the solution process considerably. At this stage many students will recognize that the number needed for the division is merely the number of ways to rearrange the 2 or 3 things chosen. Therefore, in problem 7, the 2 letters can be rearranged in 2×1, or 2!, ways. In problem 8, the 3 letters can be rearranged in $3 \times 2 \times 1$, or 3!, ways.

Students need to practice this new procedure with exercises that have sets small enough to allow the new procedure to be verified with actual lists of combinations. As students become confident in understanding this solution procedure, problems can be introduced in which the numbers are too large to permit listing the combinations, or in which extra conditions are placed on certain positions.

Finally, students will be able to return to the story about the Michigan lottery that begins this section.

Problem 9. In the Michigan lottery, the ticket purchaser chooses 6 whole numbers, trying to match the 6 numbers to be picked from the 45 balls numbered 1 through 45. How many combinations of 6 numbers can be chosen?

Solution. There are 6 slots to be filled, the first with 45 choices, the second with 44 choices, and so on. Before removing those that are the same except for order, we see that $45 \times 44 \times 43 \times 42 \times 41 \times 40$, or 5 864 443 200, arrangements of the 6 numbers are possible. There are 6!, or 720, ways to arrange 6 things. Each of the original combinations fits into a subgroup of 720 combinations that are the same except for order. Dividing the original number of arrangements by 720 produces 8 145 060 possible combinations of the 6 numbers. *(Ask students whether at $1 a ticket this lottery is a good bet when the jackpot begins at $1.5 million.)*

CONCLUSION

The development of permutations and combinations in this chapter is an example of teaching *through* problem solving, that is, using problem situations and problem-solving strategies to develop specific mathematical content. The solution procedures become the standard procedures students learn for future use. It is hoped that these explorations have been enjoyable and that readers will be willing to try some of these problems with their students.

9

Discrete Mathematics:
An Exciting and Necessary Addition
to the Secondary School Curriculum

Eric W. Hart

D ISCRETE mathematics deals with discrete objects and finite processes. Discrete mathematics topics that are already in most secondary school curricula, although they may not necessarily be taught, include matrices, sets, discrete functions and relations, permutations and combinations, discrete probability and statistics, logic, induction, and sequences. Additional topics that ought to be part of contemporary mathematics programs (NCTM 1989, 1990) include graph theory, difference equations, recursion, algorithms, linear programming, and a richer treatment of the topics in the first list, in particular, matrices and counting techniques. Many other topics in discrete mathematics are likely to be unfamiliar at the secondary school level but are widely used in business and industry and deserve consideration. These topics include game theory, Markov chains, voting theory, bin packing, apportionment, coding theory, scheduling, and fair division. Many of these topics will be discussed in more detail later in this and other chapters.

A common problem-solving method is used throughout the diverse topics of discrete mathematics. This method can be described as algorithmic problem solving, whereby one solves a problem by specifying and analyzing an algorithm that constructs the solution. It is this algorithmic method, often involving recursion, that is the unifying theme of discrete mathematics.

The algorithmic method is in contrast with, or rather complementary to, the existential method that is the hallmark of traditional mathematics, such as classical algebra, topology, and analysis. As an example, consider an important theorem in algebra: the remainder theorem. This theorem states that the remainder of a polynomial $p(x)$ on division by $(x - a)$ is $p(a)$. The traditional proof of the theorem begins by assuming that the quotient $q(x)$ and the remainder R exist. Then, by the division algorithm, $p(x) = (x - a)q(x) + R$; thus, $p(a) = (a - a)q(a) + R = 0 + R = R$. This proof seems magical to students. It is established that $p(a) = R$, but no division is actually carried out, and neither $p(a)$ nor R is computed. In contrast, an algorithmic proof consists of actually dividing the arbitrary polynomial $p(x)$ by $(x - a)$ using a systematic procedure like synthetic division. Therefore, the remainder R is actually constructed, and it is then directly observed to be $p(a)$. (See Maurer [1984] for a more complete discussion of this example.) The point of the example is to illustrate two different approaches to mathematics. It is the algorithmic approach that is characteristic of most discrete mathematics.

WHAT DISCRETE MATHEMATICS IS NOT

Discrete mathematics is not classical continuous mathematics, like calculus. One might rephrase the dichotomy between discrete and continuous by saying that discrete mathematics is concerned with countable sets, such as the integers or the rationals, while continuous mathematics is concerned with uncountable sets, such as the reals.

At first glance, discrete mathematics looks very similar to finite mathematics. Although the two do share some topics, they are definitely different. For example, topics such as graph theory and difference equations are not usually found in finite mathematics courses, discrete mathematics is not a "terminal" course as finite mathematics courses often are, and discrete mathematics places more emphasis on algorithmic methods. Differences between these courses are elaborated further in chapter 10.

With all the talk about algorithms it may look as though discrete mathematics is taking us back to the emphasis on algorithms that characterized the back-to-basics movement of the 1970s. However, there is virtually no similarity. Back-to-basics emphasized merely *performing* algorithms, and by hand at that. Discrete mathematics is concerned with designing, using, and analyzing algorithms that solve problems and develop theory, often making use of computer technology. The algorithmic approach is discussed in chapter 24.

What about the new math of the 1960s? Some common topics are present in discrete mathematics and the new math, such as sets, logic, and different number bases, but the similarity emphatically ends there. One major difference is that algorithmic thinking was essentially nonexistent in the new math. One might even distinguish between the two on the basis of practical versus theoretical. For example, one finds different number bases in both, but the new math used different bases for theoretical reasons—to firmly ground and generalize the concept of place value, whereas discrete mathematics deals with bases like 2 and 16 for the practical reason that computers use them.

Computers keep entering our discussion of discrete mathematics. In fact, discrete mathematics is the mathematical foundation of computer science, and it has been taught in college computer science departments for years. But discrete mathematics should not be tied only to computer science. It is useful in many other areas, including social science and management science as well as mathematics itself, and it differs from the typical computer science course. Some discrete mathematics topics, such as difference equations, may not be of direct interest to computer science, and certain aspects of computer science, such as proficiency in programming languages, are not strictly necessary in discrete mathematics.

A BRIEF TOUR OF DISCRETE MATHEMATICS

This section includes a sample of several discrete mathematics topics: graph theory, combinatorics, difference equations, matrices, bin packing, and voting theory. Some of these topics are treated in more depth in subsequent chapters in this yearbook.

Graph Theory

Graph theory is (very roughly) the mathematics of dots and lines. The simple idea of looking at vertices and their connections (edges) has given rise to a rich and powerful branch of mathematics. Graph theory is discussed in more detail in chapters 11 and 12 of this yearbook, so just two brief examples are given here.

1. Can you draw the house in figure 9.1 without lifting your pencil or retracing any lines? The solution of this well-known puzzle has its basis in graph theory. The house is a graph—a collection of vertices with some connecting edges—and the problem is to find an Eulerian path, that is, a path through the graph that uses each edge exactly once.

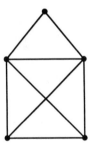

Fig. 9.1

2. Suppose that the student government at your school has five committees, and some students belong to more than one committee. How can the five committee meetings be scheduled so that the number of meeting times is kept to a minimum, and yet no two committees that share a member will be scheduled to meet at the same time? This problem can be modeled by a graph.

Let each committee be a vertex. Connect two vertices by an edge if and only if the committees they represent share a member. Then assign a color to each vertex using a different color for each meeting time. Thus, two vertices with the same color represent two meetings scheduled concurrently.

The problem now becomes one of finding the minimum number of colors needed to color the vertices of the graph in such a way that adjacent vertices, that is, vertices connected by an edge, have different colors.

The solution requires an algorithm that will produce the minimal coloring. This is an example of algorithmic problem solving. Unfortunately, no one knows an efficient algorithm that will solve this type of problem in general, but certain algorithms that can be easily taught and used will work well in most cases.

Combinatorics

Combinatorics is the mathematics of counting. More generally, it is concerned with problems that involve a finite number of possibilities, and it attempts to answer one or more of the following questions:

- Does a solution *exist*?
- *How many* solutions are there?
- Is there an *optimum* solution?

Many problems in graph theory relate to combinatorics. For instance, the graph-coloring example above is also a combinatorics problem, since the goal is to find the minimum, or optimum, number of colors needed. Another example is the problem of finding a path between two vertices in a graph. Here the three crucial questions are these:

- Does a path exist?
- How many paths are there?
- Is there an optimum path?

The most familiar combinatoric problems are those dealing with combinations and permutations.

Less familiar, yet also important, topics in combinatorics are partitions and the inclusion/exclusion principle. Finding all the different ways to write 7 as a sum of positive integers is a partition problem. Two partitions of 7 are $2 + 2 + 2 + 1$ and $4 + 3$. The inclusion/exclusion principle is actually a theorem that gives a method for counting the number of distinct elements in a finite union of sets. The method is to sum the number of elements in each individual set (which will result in counting some elements more than once); subtract the sum of the number of elements in each pair-wise intersection; add the sum of the number of elements in each three-way intersection; subtract the sum of the number of elements in each four-way intersection, and so on. This counting method can be used to solve a variety of practical problems as well as purely mathematical problems, such as counting how many integers between 1 and 3700 are divisible by 2, 3, 5, or 7. Chapters 15, 16, and 17 of this yearbook feature counting methods.

Difference Equations

A difference equation is an equation involving recursion. Recursion is the process of defining something in terms of itself in a spiral rather than a circular fashion. For example, the amount of money in an interest-paying bank account this year depends on how much was in the account last year, which depends on the amount that was in the account the previous year, and so on, until the spiral stops at the initial deposit made at time zero. The equation that describes this situation is $A(n) = (1 + r)A(n - 1)$, where r is the annual interest rate compounded annually, $A(n)$ is the amount in the account after n years, and $A(n - 1)$ is the amount after $(n - 1)$ years. We also need to include the initial deposit to indicate the starting point for the whole process, say $A(0) = \$100$. The equation is called a difference equation because of the recursion involved, and $A(0)$ is called the initial condition. This example shows that difference equations are not something new; they are just being studied now more explicitly. Note that the terms *recurrence relation* or *recurrence equation* are sometimes used instead of *difference equation*.

Difference equations can be classified in a manner analogous to differential equations. For example, the equation $A(n) = (1 + r)A(n - 1)$ is called a "first-order homogeneous linear equation." *First order* refers to the fact that $A(n)$ depends only on $A(n - 1)$ and not on any further terms like $A(n - 2)$. *Homogeneous* and *linear* mean that the right-hand side is a constant multiple of $A(n - 1)$ with no other terms.

A solution to a difference equation is a *function $A(k)$* that satisfies the equation. Sometimes, but not always, there is a "closed form" formula for the solution. Considering the difference equation above, $A(n) = (1 + r)A(n - 1)$, the closed-form solution is $A(k) = A(0)(1 + r)^k$. This solution can be found by computing $A(n)$ for a few values of n and looking for a pattern or by using a specific technique keyed to the classification of the equation. For many difference equations, closed-form solutions simply do not exist, yet they can still be fruitfully analyzed.

Difference equations are important because they describe change, and change is an essential characteristic of the real world. The standard tools used to describe continuous change are calculus and differential equations. Difference equations can be thought of as discrete analogs of differential equations. Roughly, differential equations involve slopes of tangent lines, that is, derivatives, and difference equations involve slopes of secant lines. One of the beauties of difference equations is that they allow one to tap into the vast wealth of applications of differential equations without having to wait until after learning calculus. Interesting possibilities are enumerated in chapters 18–23 of this volume.

It may look as though difference equations are only for advanced high school classes, but that is not so. Careful attention to content differentiation and the use of calculators with replay capability allow this topic to be treated in a meaningful and dynamic way with all high school students (Hirsch and Schoen 1989). The method of finite differences is appropriate for grades 7–12 (see Seymour and Shedd [1973] or Hart, Maltas, and Rich [1990]).

Matrices

The topic of matrices is one of the most overlooked topics in the traditional curriculum. Matrices are very powerful tools used in many applications. For instance, matrices can be used to store data, represent graphs, represent transformations such as rotations in the plane, solve systems of linear equations, describe Markov chains, and model games. Some of these applications are discussed in chapters 13 and 14 of this yearbook. All too often matrices are used in the classroom only to solve systems of linear equations. Sometimes they are not used at all. More instruction on matrices is probably the quickest and easiest way to get more discrete mathematics into the schools.

Bin Packing

Bin-packing problems require finding the minimum number of "bins" into which given "weights" can be packed. A simple example will illustrate this type of problem.

> Suppose eight large crates of books are to be shipped to a new library. The crates have the following weights, in hundreds of pounds: 32, 60, 56, 40, 48, 20, 60, and 64. If the moving trucks have a capacity of 12 000 pounds each, what is the minimum number of trucks needed so that all the books can be sent at the same time?

The first step in answering this question is to model the problem mathematically. We can use a bin-packing model. Let the trucks represent bins of size 120 and the crates represent the weights to be packed into the bins. In this simple example we could find the minimum number of bins needed by inspection, but we would like to have a systematic procedure—an algorithm—for finding the solution.

Many algorithms have been devised, but none are guaranteed to efficiently find the minimum number of bins in all cases. In fact, it is thought that no such general algorithm exists, although no one knows for certain. Let's look at two reasonable algorithms—one presented in a textbook (COMAP 1988), and one that was suggested by students who had a minimal background in mathematics.

The first algorithm would put the next weight to be packed into the open bin with the most space available. If it will not fit into any open bin, then open a new bin. If we pack the weights in the order in which they are listed above, then the result is as in figure 9.2. Note that five bins are needed.

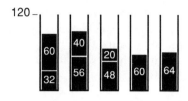

Fig. 9.2

The algorithm that the students suggested would put the next weight to be packed into the open bin with the *least* space available. This yields the result in figure 9.3.

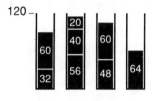

Fig. 9.3

The students were delighted when they saw that their algorithm gave a better solution than the text's algorithm. Of course this will not always happen and neither algorithm is optimal, but the process of solving this problem left the students feeling mathematically empowered.

Voting Theory

Voting theory is the mathematics of voting. One problem is to find the best possible voting scheme for a given situation. Many voting schemes are possible, such as the familiar majority and plurality and the less familiar Borda count and Condorcet winner; but a famous theorem called Arrow's theorem asserts that no voting scheme is best in all possible situations.

Another problem in weighted voting schemes, such as shareholders in a corporation where one person can hold many votes, is to measure the actual *power* of a person's vote versus the *weight* of his or her vote. For example, suppose three stockholders in a company hold 4, 3, and 2 votes. These are the *weights* of each person's vote. But do these weights accurately reflect the true *power* of a person's vote? This question can be answered as follows:

Using majority rule, 5 votes are needed to carry any proposal. Thus, a coalition must be formed by at least two people to win any vote. How many winning coalitions are there? The winning coalitions are [4, 3], [4, 2], [3, 2], and [4, 3, 2]. A sensible measure of the *power* of each person's vote is the number of winning coalitions in which his or her vote is crucial. For each person, count the number of winning coalitions in which a change of vote by that person will turn it into a losing coalition. This count is called the Banzhaf Power Index, after a lawyer who brought many inequitable voting cases to court. Thus, the stockholder with 4 votes has a Banzhaf Power Index of 2 because eliminating the 4 from [4, 3] and [4, 2] turns these coalitions into losers, but this is not so for the other coalitions. Students can easily calculate that the Banzhaf Power Index of the stockholders with 3 votes and 2 votes is also 2. So, the person with 4 votes has twice the voting weight of the person with 2 votes, yet they have the *same* voting power!

SHOULD WE TEACH
DISCRETE MATHEMATICS IN GRADES 7–12?

Every recent national report on mathematics education has included a recommendation to teach discrete mathematics in secondary schools. In particular, the NCTM's *Curriculum and Evaluation Standards for School Mathematics* (1989) recommends that the study of discrete mathematics be included for *all* students, not just college-bound students. Most recent state reports make a similar recommendation.

These recommendations are being made for many reasons:

1. *Mathematics is alive.* Students tend to believe that mathematics is dry and dusty and that all of it has been known for hundreds, or even thousands, of years. I have even heard students wondering how anyone can get a Ph.D. in mathematics—not because it would be too difficult, but because a Ph.D. requires something new, and there is nothing new left to do in mathematics. It is difficult to dispel this misconception in the context of arithmetic, algebra, geometry, and calculus, since these are indeed very old branches of mathematics, and new developments tend to be too technical to present in secondary schools. Discrete mathematics, however, is bursting with new developments and unsolved problems that can be presented to students who have very little mathematical background.

For example, the Traveling Salesman problem is easily understood as trying to find the optimum circuit through a group of cities; yet mathematicians have been working on it for decades, and no one knows an efficient solution. The bin-packing problem discussed previously is another example of this type, in which no efficient algorithm is known that will solve the problem in general. In the area of difference equations, the behavior of even a simple difference equation like $A(n + 1) = rA(n)[1 - A(n)]$ is still being studied and has implications for the understanding of chaos. (See chapter 23 of this yearbook.) Students can even do "research" on the behavior of an equation like this simply by picking different values for $A(0)$

and r and using a calculator to see what happens. Thus, one important contribution of discrete mathematics to secondary school mathematics education is to help bring the excitement and vitality of mathematics to the classroom.

2. *Problem solving and modeling are important.* Problem solving was the central theme of mathematics education in the 1980s. It is still important. The study of discrete mathematics can uniquely and significantly improve the problem-solving ability of students by developing their ability to use the powerful tool of algorithmic problem solving.

Mathematical modeling is an important and related skill that students must acquire. Yet the only modeling that many students see, if they even recognize it as such, is the representation of a word problem by an algebraic equation. Discrete mathematics provides a wealth of new and powerful mathematical models. Graphs and matrices in particular can be used to model many interesting problems and can provide a new arena for teaching, learning, and appreciating mathematical modeling.

3. *Discrete mathematics has many applications.* Discrete mathematics is used extensively in business, industry, and government. For example, graph theory can be used to schedule all the subtasks in a large project to achieve optimum completion time, difference equations are an essential mathematical tool for high-technology engineering firms, and matrices are indispensable for computer graphics.

One indication of the applicability of discrete mathematics in our modern society is shown by a major project undertaken by the Consortium for Mathematics and Its Applications (COMAP) in the 1980s. The purpose of the project was to "bring to the student the excitement of contemporary mathematical thinking and new applications." The textbook that was produced (COMAP 1988) is about 80 percent discrete mathematics. Therefore, a major project emphasizing applications of mathematics produced a textbook that is predominately discrete mathematics.

None of this is to say that other mathematics is not as useful as discrete mathematics. Rather, the point to be made is that discrete mathematics *is* broadly useful, and if we want to give a fair representation of mathematics and its applications, we should teach discrete mathematics along with the more traditional topics. If you are a classroom teacher looking for new applications, try discrete mathematics.

4. *Discrete mathematics complements and enriches the traditional curriculum.* Discrete mathematics is not a competitor of the traditional curriculum, nor is it a revolution that will radically change the way we teach or the textbooks from which we teach. It simply broadens and enriches the mathematics curriculum. In fact, discrete mathematics actually complements the traditional secondary topics of algebra, geometry, and calculus:

- Newton's method is a difference equation that was developed using calculus to find the zeros of continuous functions. It is an example of the close connection between continuous and discrete mathematics.
- Algebraic skills are needed and reinforced throughout discrete mathematics. For example, one often factors quadratics when solving second-order linear difference equations, and solving systems of linear equations is an essential part of solving problems in game theory and linear programming.
- In geometry, graph theory can be used to enrich the study of polygons and polyhedra, and difference equations give rise to the fascinating new geometry of fractals.
- Discrete mathematics can enrich basic mathematics. For example, the topic of apportionment is a good application of rounding fractions.

Although the details of these examples are not given, the point they make is that discrete mathematics is indeed complementary to the traditional secondary school curriculum.

HOW TO FIT DISCRETE MATHEMATICS INTO THE CURRICULUM

Despite an already crowded curriculum, there are several workable strategies for including discrete mathematics in the curriculum:

1. *Emphasize discrete mathematics topics that are already in place.* Matrices, counting, induction, sequences, sets, and logic are discrete mathematics topics that are already part of the present curriculum. Begin teaching discrete mathematics by emphasizing these topics and bringing them out of the obscurity into which they may have fallen.

2. *Take a discrete approach to old topics.* For example, solve systems of linear equations using matrices, represent relations using graphs and matrices, use recursive formulas to represent sequences, or use matrices to represent transformations in a unit on transformational geometry.

3. *Teach short (2-to-10-day) units on new discrete mathematics topics.* Topics like graph theory, difference equations, game theory, or linear programming can be taught. Is there room in the curriculum? Yes. Many teachers are already teaching short units without eliminating any other topics by using specific times during the year more efficiently. Also, one can make room for discrete mathematics by reducing the time spent on topics like factoring and two-column proofs, as recommended in the NCTM *Curriculum and Evaluation Standards*.

4. *Teach a full semester course on discrete mathematics.* Discrete mathematics is appropriate and useful for *all* students, so a discrete mathematics

course could fit in many places. It could be taught at the general math level, it could be a course for students who complete algebra 1 or geometry but are still not ready for algebra 2, it could be taught in the same year with precalculus, or it could be a course to follow or replace calculus. All these courses are possible, since there are many topics within discrete mathematics, and they can be approached at varying levels of sophistication.

5. *Integrate discrete mathematics into all courses.* This is part of the general integrated-curriculum issue, but it is less thorny, since discrete mathematics can be integrated separately into existing courses. Such integration can be accomplished as in the first three categories above, as well as by using discrete mathematics topics as brief examples and applications within the existing curriculum. For instance, a game theory example could be used as an application of solving systems of linear equations. An apportionment example, like apportioning seats in a state legislature, could be used as an application of rounding fractions. Computing the Banzhaf Power Index could be used as an application related to sets and subsets.

SUMMARY

This chapter has offered a brief introduction to discrete mathematics. The content and method of discrete mathematics have been examined, comparisons have been made with other trends and types of mathematics, and reasons for including discrete mathematics in the secondary school curriculum along with strategies for doing so were discussed. The overall conclusion can only be that discrete mathematics is an exciting and necessary addition to the secondary school curriculum; so let's teach it and enjoy it!

REFERENCES

Consortium for Mathematics and Its Applications (COMAP). *For All Practical Purposes: Introduction to Contemporary Mathematics.* Edited by Lynn A. Steen. New York: W. H. Freeman & Co., 1988.

Hart, Eric W., James Maltas, and Beverly Rich. "Teaching Discrete Mathematics in Grades 7–12." *Mathematics Teacher* 83 (May 1990): 362–67.

Hirsch, Christian R., and Harold L. Schoen. "A Core Curriculum for Grades 9–12." *Mathematics Teacher* 82 (December 1989): 696–701.

Maurer, Stephen B. "Two Meanings of Algorithmic Mathematics." *Mathematics Teacher* 77 (September 1984): 430–35.

National Council of Teachers of Mathematics. *Curriculum and Evaluation Standards for School Mathematics.* Reston, Va.: The Council, 1989.

――――. *Discrete Mathematics and the Secondary Mathematics Curriculum.* Reston, Va.: The Council, 1990.

Seymour, Dale, and Margaret Shedd. *Finite Differences: A Pattern-Discovery Approach to Problem Solving.* Palo Alto, Calif.: Dale Seymour Publications, 1973.

10

The Roles of Finite and Discrete Mathematics in College and High School Mathematics

Kenneth P. Bogart

Two major efforts in recent decades have attempted to establish broad-based courses in discrete mathematics as an integral part of the freshman-sophomore college mathematics program.

Let us analyze these efforts to see what they have to offer those who believe that more discrete mathematics should be in the high school curriculum. The first effort was the finite mathematics movement, begun in the 1950s under the leadership of John Kemeny and J. Laurie Snell at Dartmouth College, who developed courses for students whose interests were in the social, biological, or management sciences. We can distinguish finite mathematics as the branch of discrete mathematics that deals with step-by-step processes having a finite number of steps. This description characterizes the subject matter mathematically, but not pedagogically. Finite mathematics, then, is an essentially terminal course that attempts to develop certain skills and an awareness of the applications of mathematics outside the traditional algebra-trig-calculus sequence. Although it increases students' mathematical breadth and awareness and may introduce new skills, it does not aim to increase students' mathematical sophistication.

The second effort was spearheaded by. Anthony Ralston, among others, and led to the Mathematical Association of America's Panel on Discrete Mathematics in the First Two Years, chaired by Martha Siegel. This committee recommended (Siegel 1986) a two-semester course covering specific subjects such as combinatorics and recurrence relations and skills such as mathematical induction, combinatorial reasoning, and writing proofs. Many institutions have adopted such a course and a wide variety of textbooks have

This work was supported in part by Office of Naval Research Contract N00014-88-K-0065.

become available. Anecdotal evidence suggests that most of these courses are small, one-semester courses and are frequently not part of the mainstream in either mathematics or computer science. This type of course is characterized pedagogically by the emphasis on increasing a student's mathematical sophistication at least as much as the student's breadth and by the emphasis on computer science as a motivating topic in the way that physics is a motivating topic in calculus.

This dichotomy between the goals and the students of discrete mathematics and those of finite mathematics courses is similar to the dichotomy we see between college preparatory students—at least those preparing to take mathematics in college—and other students in high school.

WHAT IS A FINITE MATHEMATICS COURSE?

It is clear from a perusal of textbooks that there is considerable—though not complete—agreement on the content of a finite mathematics course. The classical model for finite mathematics is defined in the book by Kemeny, Snell, and Thompson (1974). The course begins with an introduction to sets and logic, which includes truth tables, truth sets, and a brief discussion of direct and indirect logical arguments (proofs in disguise). It also covers elementary combinatorics, that is, counting principles, permutations, subsets (combinations), binomial coefficients, and the binomial theorem. It uses the counting techniques and the idea of truth sets in the study of probability, covering conditional probability, independent trials, expected values, and the concept of standard deviation, with a loose description of its role in statistical inference. The course treats matrix algebra, including inverses of matrices, and the solution of systems of linear equations. It then gives selected applications, covering the use of pivoting techniques in linear programming, the application of matrices to Markov chains, or the application of matrix algebra and probability in the theory of games.

A review of recently released textbooks suggests that two aspects of the classic model have changed. Although the books reviewed have some basic material on sets, none include any logic except, perhaps, as an appendix; and all have an optional chapter on the mathematics of finance. A third (and welcome) difference beginning to appear is the addition of material on graphs. The recommended prerequisites for these finite mathematics textbooks vary from one to two years of high school algebra and seem realistic. The texts are similar in sophistication and should be accessible to a strong first-year algebra student but would require some experience with second-year algebra for a weaker student.

This review of texts suggests that most of the NCTM curriculum standards for non-college-intending students in probability and discrete mathematics and about half the standards in statistics (NCTM 1989) could be taught from a finite mathematics textbook. Among the discrete mathematics topics recommended, the treatment of recurrence equations usually appears lightly or not at all in the study of the mathematics of finance, but see Cozzens and Porter (1987) for an exception. As for the topics recommended in the probability standard, few books discuss simulation, not all discuss the normal distribution in a meaningful way, and none discuss the chi square distribution. Most finite mathematics books treat statistics as an application of probability and are unlikely to discuss the recommended topics on statistics such as curve fitting, correlation, sampling, or the manipulation and transformation of data to a significant degree.

Finite mathematics offers important ways to use the computer to ease the pain of matrix algebra, to aid in row reduction, to simulate probabilistic processes, to add binomial probabilities in order to convince students of the truth of the central limit theorem, and in other ways. This possibility was evident in the most recent edition of the Kemeny, Snell, and Thompson (1974) book, but except for Kemeny, Kurtz, and Snell's (1985) innovative "laboratory manual" supplement, the impact of computing in finite mathematics courses seems minimal or nonexistent.

WHAT IS A DISCRETE MATHEMATICS COURSE?

Few, if any, of the books on discrete mathematics released recently cover all the subjects on the MAA committee's syllabus. Possible exceptions are recent texts by Bogart (1988a) and Maurer and Ralston (1991). Virtually all books are for one-semester courses, and too little overlap is present among them to indicate a consensus on the content of a discrete mathematics course. It is clear that graph theory and mathematical induction are core topics and that computer science is a motivating theme. In an effort to obtain a clearer picture of what college faculty members mean when they talk about a discrete mathematics course, D.C. Heath conducted a survey in 1984 of college and university mathematics and computer science departments. The survey results indicate a clear consensus on what constitutes the core of a discrete mathematics course and on its one-semester duration but found less agreement on its level of sophistication and prerequisites. In the outline that follows, the topics in italics appear in about one-half to two-thirds of the respondents' courses; the other topics would be in virtually all the courses.

Popular Survey Topics in Discrete Mathematics

Set theory and related topics: Sets, relations, functions, *equivalence and ordering relations, multisets*

Combinatorics: Permutations, combinations, *partitions, recurrence relations, inclusion and exclusion, generating functions and difference equations*

Graph theory: Graphs, digraphs, trees

Mathematical induction: Induction, *recursion*

Basic logic: Truth tables, propositions, *Boolean algebra, predicate logic*

Probability/Statistics: Probability, *expected value, random variables,* binomial probabilities, *standard deviation*

Matrix algebra

The survey generally supports the recommendations of the MAA panel, but certain differences exist between the survey results and the course of study recommended by the panel. The survey did not ask about topics in number systems, and answers to open-ended questions do not suggest that such topics are included. The survey results show considerable sentiment for discussing the nature of proof in an integrated fashion rather than as a separate unit. However, the survey indicates that most of the panel's recommendations on formal logic and proof appear in many discrete mathematics courses. The panel recommends that the majority of the material in an algorithmic linear algebra unit be matrix algebra; the survey shows that a minority, close to half, of the courses have some of this material. The survey did not ask about linear programming; answers to open-ended questions suggest that this topic may appear in a significant minority of discrete mathematics courses that include linear algebra. One major point of difference between the panel recommendations and the survey results is in the panel's unit on algebraic structures. The survey suggests that Boolean algebra is the only topic in that unit likely to appear in a significant number of courses. It is a corollary of these remarks that virtually all the NCTM curriculum recommendations for discrete mathematics for college-intending students (NCTM 1989) will be met by a significant portion of, but not all, college-level discrete mathematics courses.

The earlier remarks about the value of computing exercises in finite mathematics are even more appropriate to discrete mathematics. It is useful to have students write programs to illustrate various computer algorithms discussed in the course. The Heath survey results indicate that very little programming is present in the course, although some nonprogramming software supplements illustrate various ideas in the course. With the exception of the laboratory manual (Bogart 1988b) for writing structured BASIC programs to illustrate topics in the course and the textbook (Marcus 1983) with integrated BASIC programming, no materials seem to be available for teaching a computing supplement.

DISTINGUISHING BETWEEN FINITE AND DISCRETE MATHEMATICS

Finite mathematics courses concentrate on end-user topics such as linear programming, statistics, finance, and applications of probability, whereas discrete mathematics courses concentrate equally on future-use topics like equivalence relations, induction, recursion, analysis of algorithms, and the idea of proof. This dichotomy is similar to the NCTM distinction in its *Curriculum and Evaluation Standards for School Mathematics* (NCTM 1989) between topics for all students and topics for college-intending students. It appears, however, that the NCTM recommendations for linear programming and algorithm analysis in discrete mathematics are exceptions to this rule. Linear programming, as an end-user topic that requires only straightforward algebra, might better be included for all students. Asymptotic algorithm analysis requires ideas similar to the ideas of limits and is valuable primarily to students who will use it in later study in computer science or in understanding limits. Thus, this topic might better be restricted to college-intending students. The similarities between the finite and discrete mathematics courses and the NCTM's recommended standards suggest that with appropriate modification these courses might be well suited to the secondary schools.

IS FINITE MATHEMATICS APPROPRIATE FOR HIGH SCHOOLS?

Many of the topics in finite mathematics are second-year algebra topics. Some of these topics are linear equations and inequalities of linear programming, systems of linear equations, logarithms that arise in the mathematics of finance, counting problems, probability, and matrices. Interweaving topics from algebra 2 and finite mathematics into a common one-year course could be successful. One could leave out the algebra 2 topics that are clearly preparation for calculus—factoring high-degree polynomials, the introduction to trigonometry, most of the formalism of functions and relations, even complex numbers if necessary—in order to do some of the interesting applications, such as linear programming and Markov chains (at least to the extent of Gambler's Ruin), and to have time for the more substantial (than is typical in algebra 2) matrix algebra and probability/ statistics that finite mathematics books have. The resulting course would broaden students' mathematical horizons and show them more about how mathematics is actually used. Such a course would be an appropriate substitute for algebra 2 for a student unlikely to continue studying mathematics through calculus. As noted in the earlier discussion of textbooks, the course would have the potential to cover most of the NCTM standards in probability and discrete mathematics for non-college-intending students.

The typical finite mathematics book has about forty to forty-five lessons, a number appropriate for a one-semester college course meeting three days a week. The obvious way to adapt these books to high school use is to spend more than one day on each lesson, but the homework exercises may not be quite varied enough and might not contain enough routine exercises to support this approach. Using both an algebra book and a finite mathematics book should suffice for experimental courses, and integrated finite mathematics and algebra books would be the best way to teach such a course.

IS DISCRETE MATHEMATICS APPROPRIATE FOR HIGH SCHOOLS?

The official prerequisite for most freshman-level discrete mathematics books is high school algebra, but these books are much more sophisticated in their approach to mathematics than finite mathematics books, and they aim to develop considerable sophistication in the student. A course based on one of these books would be less demanding than an Advanced Placement calculus course but more demanding than a second-year algebra course. Experience with college students suggests that all but the truly exceptional college-preparatory students would be well served by taking discrete mathematics in place of Advanced Placement calculus. This is because discrete mathematics courses place more emphasis on teaching the student to think mathematically and less emphasis on specific computational skills that are rarely, if ever, used. The top AP calculus students seen by college professors have learned both calculus skills and the ability to think mathematically; the second skill will serve them much better in college mathematics courses than the first, but the first will be more useful for the advanced-placement test. College-preparatory students with a solid background from their second-year algebra course should be ready for a course in discrete mathematics. Students who have had a precalculus course would be ready for such a course and able to skip parts of most discrete mathematics books (e.g., material on Cartesian graphs of functions, the binomial theorem, and so on). A large intersection is present between discrete mathematics and the algebra-precalculus sequence. This intersection includes induction, logarithmic and exponential functions, functions and relations, and matrices. Thus, one easy way to bring discrete mathematics into the curriculum would be to have a semester of discrete mathematics followed by a semester of precalculus mathematics (mostly trigonometry and analytic geometry, since they are not part of a discrete mathematics course). For seniors whose junior-year course was an algebra-trig course with substantial trigonometry or a precalculus course, a full year of discrete mathematics would be good preparation for college mathematics.

Note that as the college-level discrete mathematics courses are now designed, computer science is emphasized. Students don't need to know anything about computing or computer science when they enter the course, but they will know some computer science when they finish. In fact, they might understand the difference between computer programming and computer science. Although this would be a big step in the direction of computer literacy, it is not certain that this emphasis is the right one for all college-preparatory students. Thus, some redesign of discrete mathematics materials to give a more thorough mix of applications, reemphasizing some of the finite mathematics applications, would be appropriate.

Be careful about calculus preparation when trying these ideas. College calculus courses assume students can factor polynomials; solve polynomial equations; manipulate fractional expressions and functional expressions; recognize equations of circles, parabolas, and ellipses; reason intuitively about the relationship among velocity, acceleration, and distance; and visualize functions by their graphs. Whether we like it or not, college mathematics means calculus to most students. As long as business, engineering, and medical programs continue to insist on calculus for admission, college mathematics will continue to mean calculus more than anything else. In our zeal to make high school mathematics more interesting and relevant, we must be careful not to edit out the algebraic skills needed for calculus as it is presently taught.

In the long run, blending the typical functions and graphs precalculus course with a typical discrete mathematics course to yield a "Discrete and Continuous Functions and Graphs" course would produce the most appropriate course for high school students. This blend will require more than putting the textbooks for the two separate courses we are blending side by side. For example, in discrete mathematics, bijections (one-to-one and onto functions) are important because we establish a bijection between a set A and a set of integers when we count A's elements. In continuous mathematics, the fact that a function is one-to-one often means we use it to solve equations, as with solving $2^x = 3$ by using a logarithm function. An ideal course would discuss both these applications when teaching one-to-one functions. Similarly, when covering $x - y$ graphs of functions, the course could also cover the important idea of relative order of growth, discussing the so-called big-Oh notation and its relatives. It is important to the computer scientist that the function $f(x) = x \log_2(x)$ grows much more slowly than the function $g(x) = \dfrac{x^2 - x}{2}$. Each of these functions could arise in analyzing the number of seconds required by a standard method to sort a number, x, of words into alphabetical order. Although the functions are equal when x

is 4, the quadratic function grows much faster. For instance, when x is 1 000, the quadratic function is about a hundred times larger than the other one; if x is 1 000 000, the factor is more like 50 000. You can see this behavior by drawing graphs to the right scale. Organizing a combined course so that you can discuss this application when you are graphing functions involving the logarithm will take careful planning. This kind of interaction between traditional precalculus topics and discrete mathematics topics is common.

Another benefit of a combined course is that parts of discrete mathematics make an excellent training ground for calculus. Difference or recurrence equations arise from all sorts of discrete mathematical modeling problems— from computer algorithms to finance. In fact, it is possible to organize an entire discrete mathematics course around difference equations (see Sandefur 1990). The techniques used for solving difference equations are good preparation for solving differential equations. The fundamental difference is that you prove your solution is correct by mathematical induction in difference equations rather than by the fundamental theorem of calculus. It is appropriate to view mathematical induction as the foundation of a discrete mathematics course much the same way that the fundamental theorem of calculus is the foundation for a calculus course.

The discrete mathematics textbooks available vary in sophistication. Many are designed for a three-semester-hour course, with forty to fifty lessons. Although the strategy of spending two days on each lesson might adapt these books to high school use, an instructor planning to do so should carefully examine the exercises available in the book. Since the course has existed only a short time, no generally accepted collection of drill problems is available to confirm the new ideas. As a result, some of the books have some rather senseless drill problems, whereas others have no drill problems at all. This concern is equally valid for colleges and junior colleges contemplating a one-year discrete mathematics course for students with weak mathematics backgrounds. To scale a book to fit the needs of different groups of students and to fit one- or two-semester courses, divide each of the forty to fifty sections into parts, with separate exercises for each part. Key some exercises to the book, but include some less-routine problems for more sophisticated courses.

The course discussed here would cover all the topics in probability and discrete mathematics recommended in the NCTM standards for probability and discrete mathematics except, perhaps, the chi square distribution and linear programming.

CONCLUSIONS

Either of the two kinds of discrete mathematics courses now taught in colleges could be appropriate for high school students. Because finite mathematics books are aimed at the college students who took the least possible amount of mathematics in high school, these books can form the basis for a course meant for students not going on to college mathematics. Except for the difference in pace between high school and college mathematics, it should not be difficult to integrate some of the current finite mathematics books into such an algebra course. Where non-college-intending students take courses beyond the second year of algebra, finite mathematics should be seriously considered to follow the last year of algebra.

The material of the college courses in discrete mathematics is more appropriate for the precollege student than calculus. Integrating this material into a "Discrete and Continuous Functions and Graphs" course could provide a better foundation for calculus and a better foundation for college mathematics in general than the present system. For all but the best students, a year-long discrete mathematics course would be a better preparation for college mathematics than an AP calculus course.

REFERENCES

Bogart, Kenneth P. *Discrete Mathematics.* Lexington, Mass.: D.C. Heath & Co., 1988a.

_____. *Introductory Programming for Discrete Mathematics.* Lexingon, Mass.: D.C. Heath & Co., 1988b.

Cozzens, Margaret B., and Richard D. Porter. *Mathematics and Its Applications.* Lexington, Mass.: D.C. Heath & Co., 1987.

Kemeny, John G., Thomas Kurtz, and J. L. Snell. *Computing for a Course in Finite Mathematics.* Reading, Mass.: Addison-Wesley Publishing Co., 1985.

Kemeny, John G., J. L. Snell, and Gerald Thompson. *Introduction to Finite Mathematics.* 3d ed. Englewood Cliffs, N.J.: Prentice Hall, 1974.

Marcus, Marvin. *Discrete Mathematics: A Computational Approach Using BASIC.* Rockville, Md.: Computer Science Press, 1983.

Maurer, Steven, and Anthony Ralston. *Discrete Algorithmic Mathematics.* Reading, Mass.: Addison-Wesley Publishing Co., 1991.

National Council of Teachers of Mathematics. *Curriculum and Evaluation Standards for School Mathematics.* Reston, Va.: The Council, 1989.

Sandefur, James T. *Discrete Dynamical Systems: Theory and Applications.* London: Oxford University Press, 1990.

Siegel, Martha. *Report of the Panel on Discrete Mathematics in the First Two Years.* Washington, D.C.: Mathematical Association of America, 1986.

11

Graph Theory in the High School Curriculum

Robert L. Holliday

HIGH school students can be exposed to numerous areas of discrete mathematics. The NCTM *Curriculum and Evaluation Standards for School Mathematics* (NCTM 1989) identifies several mainstream topics for all students. For the strong high school junior or senior, elementary graph theory can be an important and suitable approach for presenting discrete mathematics. Graph theory is a nonthreatening topic in which an examination of standard results can give students excellent examples of the importance of precise definitions, counting arguments, inductive proofs, algorithms, and real-world applications.

THE ROLE OF DEFINITION

Many university mathematicians, when asked to describe the process of mathematics, respond with a three-word answer: definition, theorem, proof. The importance of *definition* is often lost on typical high school students. This is a problem later, in college mathematics, where students are confronted with material that contains a high density of definitions.

Graph theory also contains many definitions. But the definitions are so natural that they do not overwhelm the student. A *graph* is a finite collection of points, called *vertices*, and a finite set of *edges*. Unlike other areas of mathematics where the student is passively involved in the definition stage, graph theory gives the teacher a setting for drawing examples of graphs and asking the students to formulate a precise definition. Students will need to decide whether an edge can connect a vertex to itself, creating a *loop*, whether a vertex can be isolated, whether a pair of vertices can have *multiple edges* connecting them, or whether the edges of a graph are *directed*.

Other important concepts that arise naturally are the *degree of a vertex*, which is the number of edges that meet at a vertex, and *paths*, or *chains*, that connect two vertices by a sequence of edges. Other useful terms are *connectivity*, which requires any two vertices to be connected by a path,

87

and *circuit*, which is a path that begins and ends at the same vertex and is such that each edge in the path is traversed exactly once. In *complete graphs* every vertex is joined to every other vertex; *planar graphs* are drawn so that no two edges intersect at a nonvertex; and *simple graphs* are those with no loops or multiple edges.

The following descriptive terms are also useful: a *pendant vertex* is a vertex with exactly one edge, a *clique* is a complete subgraph, and an *isthmus* is an edge whose removal disconnects a graph. *Trees*, which are also important, are discussed later in the chapter.

The graph in figure 11.1 contains the clique *ABCD*, a pendant vertex *F*, a loop at *G*, an isolated vertex *H*, and an isthmus connecting *C* and *E*. Figure 11.2 shows a tree. Figure 11.3 has two representations of a complete graph on four vertices, which is called K_4. Although the first representation has two intersecting edges, the second representation shows that K_4 is planar.

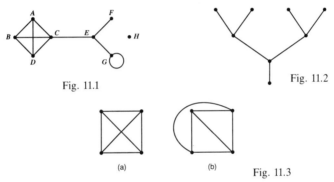

Fig. 11.1

Fig. 11.2

(a)   (b)   Fig. 11.3

COUNTING ARGUMENTS

To some mathematicians, counting arguments are the essence of discrete mathematics. These problems offer students, especially beginners, some of the best opportunities to come up with their own solutions or to understand the standard solutions. Four such problems are described in this section.

Counting Problem 1

This problem is often posed as a "handshake" problem: Prove that at any convention with a lot of handshaking, an even number of people have shaken an odd number of hands. This problem allows the student to see right away that the value of a graph comes from its modeling capabilities. Any time a relationship exists between pairs of objects in a set, it is appropriate to draw a graph to represent that relationship pictorially. In this example, the vertices of the graph are the conventioneers, and two vertices are connected by an edge if those two people shook hands. If they shook hands more than once, use multiple edges. The problem reduces to showing that a graph contains an even number of vertices with odd degree. To solve the problem, simply list (count) all the degrees: $d_1, d_2, d_3, \ldots, d_n$, where d_i denotes the degree of vertex i. Sum these degrees: $d_1 + d_2 + \ldots + d_n$. Essentially we have counted each edge exactly twice, since each edge is counted in the degree count of each of its two vertices. Thus, the addition expression above represents an even number and must contain an even number of odd summands.

Counting Problem 2

The Euler circuit problem may be the greatest counting problem of all. Certainly it is the problem that gave graph theory its start in 1736. The original question, as addressed by Euler, was this: Can the residents of Königsberg take a stroll through town (fig. 11.4), crossing each of the town's seven bridges exactly once? High school students have almost certainly seen this kind of problem posed as "Can you draw this figure without taking your pencil off the paper and without tracing over any portion of it more than once?" Clearly, both of these questions can be reduced to the question of traversing a graph, and the solution is well within the understanding of any high school student. Figure 11.5 shows the graph obtained from figure 11.4. This is again a perfect opportunity to allow students to find their own solutions. If they cannot, when they see how easy Euler's solution is, they can begin to appreciate his genius. Since this problem is thoroughly discussed in the literature, let us move on to its more troublesome "cousin."

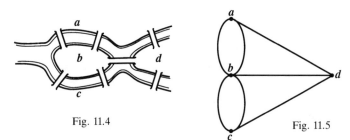

Fig. 11.4   Fig. 11.5

Counting Problem 3

The Euler circuit question requires us to traverse each *edge* of a graph exactly once, whereas the Hamiltonian circuit problem requires us to travel through a graph visiting each *vertex* exactly once. It is not necessary to travel on each edge. This problem is much more difficult than the preceding one. Euler's solution gives a technique whereby one can make a simple examination of a graph and decide whether an Euler circuit exists or not. And if one does exist, Euler explained how to find the circuit easily.

Unfortunately, no known simple technique can be used to look at a graph and decide whether a Hamiltonian circuit exists. In fact, results from theoretical computer science indicate that such a technique probably does not exist. Still, many results have been proved about Hamiltonian circuits. Many of them are phrased in this way: If a graph satisfies a certain property, then it has a Hamiltonian circuit. One such standard result is the following:

THEOREM. *Let G be a simple graph with n vertices. Suppose each vertex has degree greater than or equal to n/2. Then G has a Hamiltonian circuit.*

Remarks: First, observe that the theorem may not seem particularly surprising. If a graph has many edges (in this situation, every vertex is connected to at least half the other vertices), then it should not be a surprise that we can find a Hamiltonian circuit in the graph. Nonetheless, this is a difficult theorem for students to prove. Because graphs are pictorial, students try to prove results with a single picture instead of proving results about general graphs. In fact, the proof is not constructive in nature. It demonstrates the existence of a Hamiltonian circuit. It does not provide a recipe for finding a Hamiltonian circuit.

The standard proof of this result is found in many introductory texts. Here is an alternative proof by Donald J. Newman (Honsberger 1976). One could justify covering this theorem simply because such wonderful proofs are rare at the high school level. Newman's route is opposite to that in problem 1 above. Instead of taking a setting about people and representing it as a graph, he takes a graph with *n* vertices and rephrases the question in a "people" setting.

Proof. Suppose that the vertices of the graph represent people and that an edge indicates that those two people are friends. Then the problem of finding a Hamiltonian circuit reduces to the problem of seating all *n* people around a circular table so that each person sits between two friends. Suppose we import people into the group who are friends of everyone. Clearly, if we imported *n* such people, we could alternate them about the table with our original group and obtain a Hamiltonian seating.

Now consider a seating with a minimal number of imports. Since a Hamiltonian seating is possible with *n* imports, there is a minimal number of imports that will suffice. Call this minimal number *k*, and say figure 11.6(a) represents a minimal seating arrangement and *D* denotes an import. Then *D*'s neighbors, *A* and *B*, are clearly from the original group, since in a minimal arrangement it would be wasteful to have two imports sitting next to each other. Moreover, *A* and *B* are not friends, since if they were, we would not need *D*.

Now we make the key observation: As we scan the table in a clockwise direction, we never see the pattern "a friend of *A* immediately followed by a friend of *B*." Denote a friend of *A* by *A'* and a friend of *B* by *B'*, and suppose that such a pattern does occur, as shown in figure 11.6(b). We could then reverse the arrangement from B clockwise to A', resulting in figure 11.6(c). Everyone is still sitting between two friends, but now we don't need D anymore. This contradicts the minimality of the seating arrangement, so we can conclude that the pattern $A'B'$ does not exist. Another way of saying this is that a nonfriend of B sits to the immediate left of each friend of A.

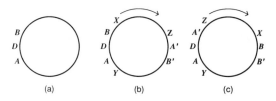

(a)   (b)   (c)

Fig. 11.6

We use a simple counting argument to complete the proof. How many friends does B have around the table? At least $n/2 + k$, since B is friends with at least $n/2$ of the original group and friends with all the imports. How many nonfriends does B have around the table? Again, at least $n/2 + k$, since this is how many friends A has and next to each friend of A is a nonfriend of B. Now, since $n + k$ people are around the table and each person is either a friend of B or a nonfriend of B, we have the following:

$$n + k \;=\; \text{number of friends of } B + \text{number of nonfriends of } B$$
$$\geq n/2 + k \qquad\qquad + n/2 + k$$
$$= n + 2k$$

From this we conclude that k must be zero. That is, we do not need any imports to obtain a seating arrangement. We can do it with the original group. Thus, the graph in question must have a Hamiltonian circuit.

Counting Problem 4

Trees provide an important link between mathematics and computer science; so any treatment of graph theory should mention a few results about trees. A tree is a connected graph with no circuits; a circuit is a path that begins and ends at the same vertex. Another useful fact, which is proved later, is that if a tree has n vertices, then it has $n-1$ edges. From these facts, we use simple counting to prove the following result: *A graph with n vertices, n − 1 edges, and no circuits is a tree.*

To prove this statement, students should keep track of several concepts. They need to recognize that there is something to be proved and that the proof does not follow directly from the definition. What is missing is connectivity. Thus, we must show that a graph with n vertices,

$n - 1$ edges, and no circuits is connected. Suppose that a graph G satisfies the hypotheses above. Split G into its connected pieces, called connected components. Call these connected components T_1, T_2, T_3, . . ., T_k, and suppose that T_1 has n_1 vertices, T_2 has n_2 vertices, . . ., and T_k has n_k vertices. We know that $n = n_1 + n_2 + \ldots + n_k$. Each of these connected components has no circuits because the original graph G does not; therefore, each of them is a tree. Now, count edges. Since each T_i is a tree, it contains $n_i - 1$ edges, and we have another equation: $n - 1 = (n_1 - 1) + (n_2 - 1) + \ldots + (n_k - 1)$. Note that this second equation contradicts the first one if more than one subtree (connected component) exists. Therefore only one connected component is present, and the original graph must have been connected.

PROOFS BY MATHEMATICAL INDUCTION

More than any other concept, mathematical induction is learned by mimicking the teacher. Most students learn induction by proving formulas, usually adding the $(n + 1)$st term to both sides of an equation. Thus, any setting that allows the opportunity for different induction proofs is healthy. Graph theory affords many opportunities. In this section, we prove one result by mathematical induction. Additional theorems in graph theory that involve mathematical induction can be found in chapters 12 and 21.

THEOREM. *A tree with n vertices has exactly n − 1 edges.*

Analysis and Proof. Before we are ready to prove this theorem, we need an auxiliary fact: A tree has at least one pendant vertex, that is, a vertex of degree 1. If this were not true, we could pick a vertex of our tree, say n_1, find a vertex connected to it, say n_2, and traverse a path through the tree: n_1 to n_2 to n_3 to. . . . How do we know there is a vertex n_3 that is connected to n_2? If no vertex is a pendant vertex, then every vertex is connected to at least two other vertices. Since a tree (and all graphs) has a finite number of vertices, we must eventually repeat a vertex as we follow the path. But this would give us a circuit, contradicting the definition of a tree. Thus, we cannot continue the path indefinitely. There is a vertex, n_k, that is not connected to another vertex, n_{k+1}. This vertex n_k is a pendant vertex.

Now that we know a pendant vertex exists, we can prove the result above by induction. The trivial case is obvious: A tree with 1 vertex has 0 edges, since the only possible edge would be a loop, which would yield a circuit. Suppose the result is true for all trees with n vertices, and consider a tree with $n + 1$ vertices. Since each tree has a pendant vertex, remove this vertex and its only edge from the tree. This leaves a graph with n vertices. Since the original graph had no circuits, surely removing an edge does not create one; so the remaining graph has no circuits. But the graph is still connected, since the edge we removed only connected the pendant vertex to the rest of the graph. Therefore, the remaining graph is a tree. By induction, this tree of n vertices has $n - 1$ edges. So the original tree of $n + 1$ vertices must have had n edges, and the result is proved.

FORMULATING ALGORITHMS

Many important problems today require algorithmic solutions. Some problems are so large that the solutions need to be found using a computer. Once we know how to solve these problems, we need to be able to express our solutions in a form that is understood by a computer. We need to formulate an algorithm that can be implemented by a computer. This section considers one such problem.

Algorithm Problem

Given a graph, color the vertices using as few colors as possible so that vertices connected with an edge are colored with different colors.

If we insist on using the smallest number of colors possible, this problem, at least for large graphs, is virtually unsolvable, even with the aid of the most powerful computers. It might take many thousands of centuries to determine this smallest number. So, we consider an approximate algorithm—one that executes quickly and uses, we hope, a relatively small number of colors, even though this number might be slightly larger than the absolute minimum. Here is a reasonable algorithm:

- *Step 1:* Arrange the vertices in order of decreasing degree. That is, the first vertex listed has the largest degree, and the last vertex listed has the smallest degree.
- *Step 2:* Choose a color (say red) and color vertex 1 red. Continue through the list of vertices (as ordered by step 1), coloring any vertex red that is allowed to be red. For example, if vertex 2 is connected to vertex 1 by an edge, it cannot be colored red. But if it is not connected to vertex 1, then it can be red.
- *Step 3:* When we reach the end of the list, choose another color (maybe blue), return to the top of the list, and color any vertex that we can with this new color.
- *Step 4:* Repeat step 3 until all vertices are colored.

This algorithm can be shown to color the graph efficiently. To see that the coloring produced might only be an approximation to the optimal coloring, consider the graph in figure 11.7.

Fig. 11.7

Four of the vertices have degree 2, and two of the vertices have degree 1. The first step of the algorithm requires that the vertices be listed in decreasing order by degree. If the algorithm breaks ties alphabetically, then vertices A and B are listed first. The algorithm colors A and B the same color. Then we must use two additional colors for vertices C and D. Yet it is easy to see, by alternating colors, that the graph can indeed be colored with two colors.

In classes where the students also know programming, a good exercise begins with a graph drawn on paper and finishes with a computer program that reads the graph and specifies how the vertices are to be colored. This opportunity to participate in the complete cycle of the modern-day problem-solving process is invaluable. See Welsh and Powell (1967) and Holliday and Carmony (1987) for a complete discussion of this problem.

APPLICATIONS

For those students who do not appreciate mathematics for its intrinsic beauty and ask for relevance to real-world applications, graph theory may be the perfect setting. Certainly, applications for other areas of mathematics can be seen as well, but not so easily. For example, we might be able to convince our students that calculus can be used to help civil engineers build better bridges, but the students still might not see how it really works. But in graph theory, we can explain the applications, the students can see how they work, and they can actually solve real problems. For example, as seen in chapter 9, coloring a graph has direct applications to scheduling situations. We mention a few additional examples in this section. Introductory textbooks are filled with many more.

Euler and Hamiltonian Circuits

The applications here are relatively obvious. Street-cleaning crews would like Euler circuits over their territories, whereas traveling salespersons would like Hamiltonian circuits over their territories. In fact, graph theory has been used to route the garbage trucks of New York City.

Trees

Trees are, in a sense, the minimal graphs for connecting a collection of vertices. Thus, they are important in any kind of network—cities on an airline schedule, telephone communication links, and so on. In chemistry, trees have been used to predict the existence and structure of various chemical compounds.

Bipartite Graphs

We have only scratched the surface of graph theory in this chapter. Numerous other directions are available in which to head, many of them within the grasp of high school students. A *bipartite* graph is one in which the vertices of the graph are divided into two sets, and each edge of the graph connects a vertex from one of the sets to a vertex in the other set.

Bipartite graphs can be used to model a situation where several job applicants list the jobs they can perform and we match job openings with the applicants so that we can fill the most positions. This problem is known in some formulations as the "marriage problem." Its proof, by Philip Hall, is another excellent example of mathematical induction.

SUMMARY

The movement toward discrete mathematics at all levels of the curriculum has occurred for many reasons. Computers and the need for algorithmic thinking are often cited as the main reasons. Just as important, if not more so, is that discrete mathematics, particularly at the formative stages of serious mathematical study, furnishes the opportunity for a student to see and actively engage in the entire mathematical process without a sophisticated mathematical background.

Graph theory, perhaps the most visual and familiar area of discrete mathematics, certainly affords the high school student every opportunity to participate actively in the mathematics process. A student who covers the topics outlined in this chapter, or similar topics from books listed in the Bibliography, can experience and understand many different proof techniques for important real-world applications.

BIBLIOGRAPHY

Brualdi, Richard. *Introductory Combinatorics.* New York: North Holland, 1977.

Holliday, Robert, and Lowell Carmony. "A Scheduling Problem: Modeling, Approximate Algorithms, and Implementation." *ACM SIGCSE Bulletin* 19 (February 1987): 473–80.

Honsberger, Ross. *Mathematical Gems II.* Dolciani Mathematical Expositions No. 2. Washington, D.C.: Mathematical Association of America, 1976.

National Council of Teachers of Mathematics. *Curriculum and Evaluation Standards for School Mathematics.* Reston, Va.: The Council, 1989.

Roberts, Fred. *Applied Combinatorics.* Englewood Cliffs, N.J.: Prentice-Hall, 1984.

Welsh, D., and M. Powell. "An Upper Bound for the Chromatic Number of a Graph and Its Application to Timetabling Problems." *Computer Journal* 10 (May 1967): 85–86.

12

Discovering and Applying Euler's Formula

Donald W. Miller

ONE of the most useful tools of graph theory is Euler's formula, which relates the numbers of vertices, edges, and faces of a graph. In many discrete mathematics texts this formula is simply announced, and verification is replaced by simply checking a number of cases. By conjecturing the formula on their own, however, students will experience the thrill of discovery—a thrill that memorization cannot match.

EULER'S FORMULA: DISCOVERY

Let us begin our approach to Euler's formula by reviewing some definitions. A *graph* consists of a finite set of points, called *vertices*, and a finite set of segments, called *edges*, such that the endpoints of each edge are members of the set of vertices. We shall restrict our attention to graphs that are *planar*, that is, graphs that can be drawn in a plane. In such a graph two edges can intersect only in a common endpoint. We shall also require that our graphs be *connected*, that is, that given any two vertices V_1 and V_2 of the graph, it is possible to move from V_1 to V_2 along (some of) the edges of the graph. Every such graph partitions the plane into regions, or *faces*, one of which is the infinite region that surrounds the entire network of vertices and edges. Several examples of such graphs are shown in figure 12.1.

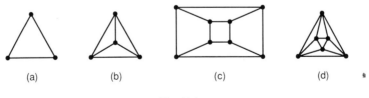

(a)   (b)   (c)   (d)

Fig. 12.1

We shall denote the number of edges, faces, and vertices in a graph by e, f, and v, respectively. For example, the graph in figure 12.1(b) satisfies $e =$

$6, f = 4$, and $v = 4$. Euler's formula is a response to the following question: Is there a relationship among e, f, and v that is valid for every graph?

We begin to explore this question by collecting data from the graphs shown in figure 12.1, leading to the following table:

Graph	e	f	v
(a)	3	2	3
(b)	6	4	4
(c)	12	6	8
(d)	12	8	6

What can be concluded about e, f, and v from this set of data?

Students, challenged by this question, might respond in several ways: "e is always the biggest." "v divides the product of e and f." "There is no connection, because f can be smaller than v, or equal to v, or bigger than v."

Suppose we try a different approach. Divide the class into four or five small groups, and give each group written instructions. Group A is asked to draw several graphs, each of which has exactly 3 faces. Groups B, C, and D are given similar instructions, but with the number of faces equal to 4, 5, and 6, respectively. No group is aware of the specific instructions given to the other groups. After a few minutes give each group a second set of (identical) instructions: "Choose three of your graphs, tabulate the number of edges and vertices for each of them, and see what your data suggest to you." Within minutes each group will have a conjecture; however, each group will probably disagree with every other group.

Now ask each group to put its graphs, data table, and conclusion on the chalkboard. Something like figures 12.2–12.4 will appear.

Confronted with these selected data—their own and that of the other groups—the students quickly see the role played by f, the number of faces.

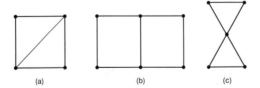

(a)   (b)   (c)

Group A:

Graph	e	v	
(a)	5	4	
(b)	7	6	
(c)	6	5	*Conclusion: $e - v = 1$.*

Fig. 12.2

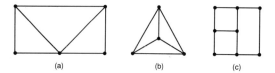

Group B:

Graph	e	v	
(a)	7	5	
(b)	6	4	
(c)	10	8	*Conclusion: e − v = 2.*

Fig. 12.3

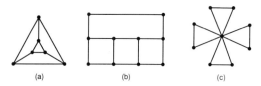

Group C:

Graph	e	v	
(a)	9	6	
(b)	13	10	
(c)	12	9	*Conclusion: e − v = 3.*

Fig. 12.4

They will almost certainly conclude that $e - v$ is equal to $f - 2$. With a slight rephrasing, then, we have the following theorem:

THEOREM (Euler's formula). *For each connected planar graph with v vertices, e edges, and f faces, $v - e + f = 2$.*

EULER'S FORMULA: PROOF

To verify Euler's formula, we consider an arbitrary graph that satisfies our definition of a graph. We proceed to modify this graph in two stages.

Stage 1

Triangulate the graph by adding line segments that (*a*) lie in the interior of one of the finite regions (faces) of the graph *and* (*b*) join two vertices that were not previously connected. Note that when such a new segment is added,

v is unchanged, e increases by 1, and f increases by 1,

so the value of the expression $v - e + f$ is unchanged. Repeat this process until every finite region of the original graph has been partitioned into triangular regions. For example, one triangulated version (and there are many) of the graph in figure 12.5(a) is shown in figure 12.5(b).

(a)

(b)

Fig. 12.5

Stage 2

Reduce the triangulated graph to a single triangle by two types of deletions.

Type (a). If there is a vertex V on the boundary of the graph at which exactly two edges, say E_1 and E_2, meet, delete V, E_1, and E_2 (but retain the other vertex on each of those edges).

Type (b). If no deletion of type (a) is possible, delete a single boundary edge of the graph but retain its endpoints.

After performing a deletion of type (b), return to type (a) deletions as long as possible before performing another type (b) deletion. For example, starting with the (triangulated) graph in figure 12.6(a), we find that the reduction process (in which there is a considerable amount of freedom) could lead to the graph in figure 12.6(e).

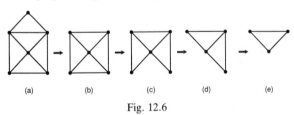
(a)   (b)   (c)   (d)   (e)

Fig. 12.6

What effect does each of these two types of deletions have on the value of the expression $v - e + f$? Under a deletion of type (a),

v decreases by 1, e decreases by 2, and f decreases by 1,

so the value of $v - e + f$ is unchanged. Under a deletion of type (b),

v is unchanged, e decreases by 1, and f decreases by 1,

so again the value of $v - e + f$ is unchanged.

But as we have seen, the original graph can be triangulated by the addition of edges (Stage 1), then reduced to a single triangle by the deletion of edges and vertices (Stage 2). Moreover, the value of the expression $v -$

$e + f$ is not changed in the process, so its final value (for a single triangle) coincides with its initial value (for the original graph). But for a graph consisting of a single triangle, $v = 3$, $e = 3$, and $f = 2$, so $v - e + f = 2$. This completes the proof.

EULER'S FORMULA: WHERE DOES IT LEAD?

For more than a century one of the most famous unsolved problems of mathematics was the four color problem: Are four colors sufficient to color every planar map in such a way that every pair of countries with a common border are colored differently? (In this problem it is assumed that each country consists of a single connected region, which rules out countries like the United States, in which Alaska and Hawaii are disconnected from the other states.)

First posed in 1852, it was believed to have been answered in 1879 (by a proof that was later shown to be invalid!). The question was not settled until 1976, when two mathematicians at the University of Illinois, Appel and Haken (1976), announced that they had proved the long-suspected result:

THEOREM. *Every planar map can be colored with four or fewer colors.*

Their proof, which required some 1200 hours of computer time to check more than 1900 cases, occupies 189 pages in the *Illinois Journal of Mathematics*. One of the early steps in the solution of the four color problem was the following result, known as the six color theorem:

THEOREM. *Every planar map can be colored with six or fewer colors.*

This result is far more accessible than the four color theorem. The proof given here, a modification of one by Rebman (1979), depends on two fundamental results of discrete mathematics: Euler's formula and the (generalized) pigeonhole principle.

The pigeonhole principle, in its simplest form, can be expressed as follows: If more than n objects are distributed into n compartments (where n is a positive integer), then some compartment will contain more than one object. The principle can be extended in the following way to the generalized pigeonhole principle: If more than tn objects are distributed into n compartments (where t and n are positive integers), then some compartment will contain more than t objects. Similarly, if fewer than tn objects are distributed into n compartments, then some compartment will contain fewer than t objects.

For example, if 1025 raffle tickets are randomly distributed among the visitors to an auto show and if the total attendance at the show is 256, then, since $1025 > 1024 = (4)(256)$, some visitor must receive more than 4 tickets. But if only 1020 tickets are distributed, some visitors will receive fewer than 4 tickets.

In proving the six color theorem, we shall modify some of our earlier language, taking the terms *graph, edge,* and *face* as synonymous with the terms *map, boundary,* and *region,* respectively. In addition, we shall interpret the word *map* to mean a finite, connected, planar map. It is convenient to define the *degree* of a vertex V of a map to be the number of edges of the map that have V as an endpoint. For example, in the map shown in figure 12.7(a), the vertices P, Q, and R have respective degrees 2, 4, and 3. We shall assume that every vertex in our map has degree ≥ 3. This assumption can be made without loss of generality; for example, the map shown in figure 12.7(a), which contains a vertex of degree 2, can be transformed into the map of figure 12.7(b), in which every vertex has degree ≥ 3, without affecting the number of colors required to color the map.

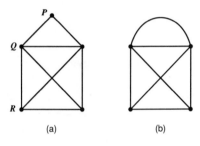

(a)   (b)

Fig. 12.7

Now suppose that we are given an arbitrary map that satisfies the conditions above. As before, we denote the number of boundaries, regions, and vertices in the map by e, f, and v, respectively. We wish to show that this map can be colored using at most six colors.

Let us think of this map as that of a small, uninhabited island on which some of the pirate Blackbeard's buried treasure has recently been discovered. The country of Caribbee, which owns the island, has subdivided it into mining claims, which are represented by the regions of the map. All these claims, together with a claim consisting of the immediately surrounding ocean, have been leased for exploration. Fences have been erected along the borders of each claim, including a fence around the perimeter of the island. At each point P where three (or more) such fences meet, a lightpost has been erected, with a powerful floodlight directed along each fence having P as an endpoint.

For example, Blackbeard's Island might look like figure 12.8 (where the floodlights are indicated by arrows).

Since at least three edges meet at each vertex of the map, there must be at least three floodlights mounted on each of the lightposts. Therefore the total number of floodlights must be at least $3v$. But since each of the e edges has exactly two floodlights directed along it (one at each of its endpoints),

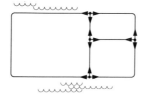

Fig. 12.8

the total number of floodlights is equal to $2e$. This proves

Lemma 1. $3v \leq 2e$.

In order to prevent the violence that might result if one explorer began to dig on the claim of another, the governor of Caribbee has directed that a flagpole be erected in each of the regions (including the single region made up of the coastal waters), that a solid-colored flag be flown on each flagpole, and that the workers on each claim wear a shirt that matches the color of their flag. The governor has also required that the colors of the flags are to be chosen in such a way that any two claims that have a common boundary must have flags of differing colors.

Thus the problem of coloring the given map is equivalent to that of coloring (i.e., selecting the colors of) the flags.

The claim holders have decided to post a guard along each of their boundaries. In addition, the holder of the ocean exploration claim has assigned a one-person boat patrol along the water boundary of each oceanside claim. Guards in the map shown in figure 12.9 are indicated by small circles.

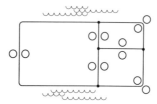

Fig. 12.9

Each of the e boundaries is then patrolled by two guards, one on each side of the fence that defines that boundary. Therefore the total number of guards patrolling the borders of the f regions is $2e$, for an average of $g = 2e/f$ guards per region. Thus $2e = fg$, so

$$e = \frac{fg}{2}. \tag{1}$$

Substituting this value of e into the inequality of lemma 1 yields the result $3v \leq fg$, or

$$v \leq \frac{fg}{3}. \tag{2}$$

Therefore $2 = v - e + f$ by Euler's formula

$$= v - \frac{fg}{2} + f \qquad \text{by (1)}$$

$$\leq \frac{fg}{3} - \frac{fg}{2} + f \qquad \text{by (2)}.$$

Thus $2 \leq f - fg/6$, that is, $fg/6 \leq f - 2$. Therefore $fg \leq 6f - 12$, so $g \leq (6f - 12)/f = 6 - 12/f < 6$. But if the average number of guards per region is less than 6, then by the generalized pigeonhole principle some region will require at most 5 guards, and hence must have at most 5 boundaries. This completes the proof of

Lemma 2. Some region of the map has at most five neighboring regions.

This lemma leads us directly to the proof of the six color theorem. The proof is by induction on f, the total number of regions of the map.

The theorem is certainly true for any value of f less than or equal to 6. Assume inductively that the theorem holds for all maps with fewer than f regions. Consider an arbitrary map containing exactly f regions; regard it as the map of Blackbeard's Island and the surrounding water, subdivided into f exploration claims.

By lemma 2, one of these f claims, say region R, has at most 5 neighboring claims. Suppose that the holder of one of those claims, say of region S, leases region R and takes down the fence that separates R from S. The map of the new system of claims then contains only $f - 1$ regions, so by our induction hypothesis, it can be colored with at most 6 colors.

Now suppose that a few months later, when the lease of region R expires, region R is reestablished as a separate claim and that the fence that once separated R from S is rebuilt. The map of this new system of claims has been colored with 6 (or fewer) colors, but region R is colored the same as at least one of its neighbors, S. With the exception of region R, every two regions that share a common boundary are colored differently.

However, since region R has at most five neighbors, there is a sixth color available to color R that differs from the colors assigned to the neighbors of R. Choosing this color for R produces the desired coloring of the given map, using at most six colors. Our theorem is proved.

REFERENCES

Appel, Kenneth, and Wolfgang Haken. "Every Planar Map Is Four Colorable." *Bulletin of the American Mathematical Society* 82 (September 1976): 711–12.

Rebman, Kenneth R. "The Pigeonhole Principle (What It Is, How It Works, and How It Applies to Map Coloring." *Two-Year College Mathematics Journal* 10 (1979): 3–12.

13

Matrices at the Secondary School Level

Denisse R. Thompson
Sharon L. Senk
Steven S. Viktora

R ECENT applications of matrices to business, computer science, and physical sciences suggest that matrices be given a more prominent role in the secondary school mathematics curriculum. In line with this need, the *Curriculum and Evaluation Standards for School Mathematics* (NCTM 1989) calls for increased attention to matrices in grades 9–12 for *all* students. This chapter illustrates important uses of matrices and discusses where in the secondary school curriculum these topics may be placed.

USES OF MATRICES

We begin by considering matrices as a tool for storing numerical data, which is a natural context for the introduction of matrix operations. We then discuss using matrices to solve linear systems. Returning to data storage, we use matrices to represent geometric figures and transformations. Building on this geometric interpretation, we give examples showing how matrices can be applied in the study of trigonometry.

Storing and Organizing Numerical Data

One of the simplest uses of matrices is to store data. In fact, this is the most common use outside mathematics. A look at any newspaper sports section reveals conference or league statistics that are written in a row-

This work was partially supported by grants to the University of Chicago School Mathematics Project from the Amoco Foundation, the Carnegie Corporation of New York, and the General Electric Foundation.

column format; that is, the statistics are written in a matrix. The table in figure 13.1 lists the batting record of each National League baseball team in 1988. Because there are 12 rows and 6 columns of data, we say the matrix has dimensions 12 × 6. (If rows and columns are permitted to contain nonnumerical data, we could say the matrix has dimensions 13 × 7.)

Club	PCT	National League 1988 Club Batting AB	R	H	HR	SB
Chicago	.261	5675	660	1481	113	120
New York	.256	5408	703	1387	152	140
Montreal	.251	5573	628	1400	107	189
St. Louis	.249	5518	578	1373	71	234
San Francisco	.248	5450	670	1353	113	121
Los Angeles	.248	5431	628	1346	99	131
San Diego	.247	5366	594	1325	94	123
Pittsburgh	.247	5379	651	1327	110	119
Cincinnati	.246	5426	641	1334	122	207
Houston	.244	5494	617	1338	96	198
Atlanta	.242	5440	555	1319	96	95
Philadelphia	.239	5403	597	1294	106	112

Source: *The World Almanac and Book of Facts, 1989*

Fig. 13.1

Questions that might be asked about the matrix in figure 13.1 are these:

1. How many runs did the Chicago Cubs score? (660)
2. What was the batting average of the Los Angeles Dodgers, the 1988 League champions? (.248)

Instead of being asked questions about a given matrix, students could be asked to find and organize data about a topic of interest to them. Although this is not a difficult task, there is room for mathematical discussion. Some students may store the data in an $n \times p$ matrix, others in a $p \times n$ matrix. Although the matrices are not equal, they are certainly equivalent representations of the data. Flexibility in presenting information is an important mathematical skill that is too seldom discussed.

Tax tables, airline schedules, and computer spreadsheets offer a variety of other situations that can be represented by matrices. Reading data from matrices and using matrices to store data are certainly accessible to students in grades 7 to 10.

Prior to the second year of algebra, the introduction of matrices does not require the use of subscripts to identify entries. Later, however, the use of subscripts may be desirable. For instance, when matrices are stored in computers, questions naturally arise about how to access the data. Consider the 3 × 4 matrix in figure 13.2. (To simplify the generalizations, the first row is labeled row 0 and the first column is labeled column 0.)

	column 0	column 1	column 2	column 3
row 0	a_{00}	a_{01}	a_{02}	a_{03}
row 1	a_{10}	a_{11}	a_{12}	a_{13}
row 2	a_{20}	a_{21}	a_{22}	a_{23}

Fig. 13.2

In a computer, this matrix is stored in a linear rather than a rectangular fashion, possibly as follows:

location number	0	1	2	3	4	5	6	7	8	9	10	11
matrix element	a_{00}	a_{01}	a_{02}	a_{03}	a_{10}	a_{11}	a_{12}	a_{13}	a_{20}	a_{21}	a_{22}	a_{23}

We say that this matrix is stored in row order because row 0 is stored before row 1, which is stored before row 2. It is possible to store a matrix in column order; then the following discussion would need appropriate modifications.

A computer programmer would be interested in knowing how the location number n is related to the subscripts i and j. An answer is provided by the division algorithm: for any two positive integers n and d, with $d < n$, there exist integers q and r such that $n = dq + r$ and $0 \leq r < d$. Storing the matrix in row order means that the elements can be considered as split into groups of d elements per row, where, in this case, $d = 4$. If the location number 9 is divided by 4, the quotient is 2 and the remainder is 1. Note that the element in location number 9 is a_{21}; the row subscript equals the quotient and the column subscript equals the remainder. In general, the element in location n of a list representing a matrix with d elements per row is a_{ij} where $n = di + j$. Of course, the division algorithm can be used in reverse to determine the location number n from d and the subscripts i and j.

Operations with Matrices

Addition and subtraction of matrices are within the grasp of students as early as grade 7. Consider the matrices that give the arrest rates of persons under 18 (per thousand 14-to-17-year-olds) and of persons 18 to 24 (per thousand 18-to-24-year-olds) by type of crime. To discuss changes in the arrest rates from 1965 to 1985, corresponding elements in the two matrices must be subtracted. The result is the matrix shown in figure 13.3.

Several questions might now be asked:

1. By how much did the arrest rate for drug abuse increase for 18-to-24-year-olds between 1965 and 1985? [9.3 per 1000]
2. For which age group and which crime has the greatest increase in arrest rate occurred? [Drunk driving for 18-to-24-year-olds]

	1985			1965					
	under 18	18 to 24		under 18	18 to 24			under 18	18 to 24
Arson	0.5	0.1		0.3	0.0			0.2	0.1
Drug Abuse	5.4	10.2		0.4	0.9			5.0	9.3
Drunk Driving	1.4	15.6	−	0.1	1.9	=		1.3	13.7
Larceny/Theft	26.0	11.7		14.9	4.1			11.1	7.6
Vandalism	0.6	2.2		4.9	6.8			−4.3	−4.6

Source: *Youth Indicators 1988: Trends in the Well-Being of American Youth*

Fig. 13.3

3. Did the arrest rate for any crime decrease for either age group between 1965 and 1985? If so, for which one(s)? [yes—vandalism for both age groups]

Matrix multiplication also has many uses within the grasp of average high school students. For instance, suppose a contractor builds four model houses—I, II, III, and IV—in three developments—Hill, Plain, and Dale. Matrix A gives the number of each model built last year, and matrix B gives the number of exterior doors and windows in each of the four models.

		Matrix A						Matrix B	
	I	II	III	IV				Doors	Windows
Hill	10	5	1	2		I		2	12
Plain	5	10	2	5		II		2	20
Dale	6	4	5	3		III		3	15
						IV		3	20

Multiplying matrix A by matrix B results in the 3×2 matrix below.

$$\begin{bmatrix} 10 & 5 & 1 & 2 \\ 5 & 10 & 2 & 5 \\ 6 & 4 & 5 & 3 \end{bmatrix} \cdot \begin{bmatrix} 2 & 12 \\ 2 & 20 \\ 3 & 15 \\ 3 & 20 \end{bmatrix} = \begin{matrix} & \text{Doors} & \text{Windows} \\ \begin{bmatrix} 39 & 275 \\ 51 & 390 \\ 44 & 287 \end{bmatrix} & \begin{matrix} \text{Hill} \\ \text{Plain} \\ \text{Dale} \end{matrix} \end{matrix}$$

Because each element in row 1 of A represents the number of homes of each model in the Hill development and each element of column 1 of B represents the number of doors in each model, the element in row 1 and column 1 of AB represents the total number of doors in the homes in the Hill development. Thus, AB gives the number of doors and windows in each housing development.

Note that the product BA cannot be calculated, because the number of columns (2) of B does not equal the number of rows (3) of A. Many students are surprised to learn that *matrix multiplication is not commutative*.

They may also be surprised to learn that *matrix multiplication is associative*. Consider again the housing contractor. Let C be the matrix giving the average unit cost in dollars of each door and window. It is instructive to calculate each of *(AB)C* and *A(BC)*.

$$\begin{matrix} & \text{Matrix } C \\ & \text{unit} \\ & \text{cost} \\ \begin{matrix} \text{door} \\ \text{window} \end{matrix} & \begin{bmatrix} 100 \\ 80 \end{bmatrix} \end{matrix}$$

(AB)C is calculated by first finding the number of doors and windows in each of the developments and then finding the total cost of windows and doors for each development.

$$\left(\begin{bmatrix} 10 & 5 & 1 & 2 \\ 5 & 10 & 2 & 5 \\ 6 & 4 & 5 & 3 \end{bmatrix} \begin{bmatrix} 2 & 12 \\ 2 & 20 \\ 3 & 15 \\ 3 & 20 \end{bmatrix} \right) \begin{bmatrix} 100 \\ 80 \end{bmatrix} = \begin{bmatrix} 39 & 275 \\ 51 & 390 \\ 44 & 287 \end{bmatrix} \begin{bmatrix} 100 \\ 80 \end{bmatrix} = \begin{bmatrix} 25900 \\ 36300 \\ 27360 \end{bmatrix}$$

Alternatively, *A(BC)* is calculated by first finding the cost of the doors and windows in each model and then finding their total cost for each development.

$$\begin{bmatrix} 10 & 5 & 1 & 2 \\ 5 & 10 & 2 & 5 \\ 6 & 4 & 5 & 3 \end{bmatrix} \left(\begin{bmatrix} 2 & 12 \\ 2 & 20 \\ 3 & 15 \\ 3 & 20 \end{bmatrix} \begin{bmatrix} 100 \\ 80 \end{bmatrix} \right) = \begin{bmatrix} 10 & 5 & 1 & 2 \\ 5 & 10 & 2 & 5 \\ 6 & 4 & 5 & 3 \end{bmatrix} \begin{bmatrix} 1160 \\ 1800 \\ 1500 \\ 1900 \end{bmatrix} = \begin{bmatrix} 25900 \\ 36300 \\ 27360 \end{bmatrix}$$

The introduction of matrix operations at an early point in the curriculum provides an opportunity for students to gain facility with matrices and to see specific instances where common properties, such as commutativity, fail to hold. Such familiarity is an important prerequisite for studying the mathematical structure of matrices.

Matrices for Graphs

Matrices can be used to store data about graphs. Here the word *graph* does not represent a set of ordered pairs in the Cartesian plane but rather a geometric figure consisting of points (vertices) and edges connecting some of these points. If the edges are assigned a direction, the graph is called *directed*. Directed graphs are useful for analyzing flowcharts or for solving problems about shortest paths.

Consider the pictorial representation in figure 13.4 of a directed graph. The number of direct routes between pairs of vertices can be represented by a connecting matrix. If R is the matrix for the number of direct routes, $r_{11} = 0$ because there is no route from A to A that goes along exactly one edge. The entry r_{12} is 1 because there is 1 direct route from A to B. The entire connecting matrix R for one-leg routes is shown here.

As illustrated in figure 13.5, there are 3 two-leg routes from A to A.

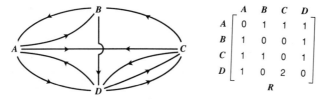

Fig. 13.4

Further, as shown in figure 13.6, there is 1 two-leg route from A to B.

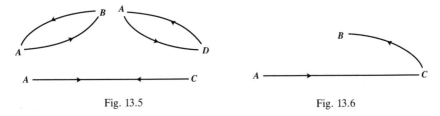

Fig. 13.5   Fig. 13.6

The matrix T, representing the number of possible two-leg routes among A, B, C, and D, is shown below.

$$\begin{bmatrix} 3 & 1 & 2 & 2 \\ 1 & 1 & 3 & 1 \\ 2 & 1 & 3 & 2 \\ 2 & 3 & 1 & 3 \end{bmatrix}$$
$$T$$

Note that any connecting matrix must be square. Thus powers of such matrices can be calculated. Interestingly,

$$R^2 = \begin{bmatrix} 0 & 1 & 1 & 1 \\ 1 & 0 & 0 & 1 \\ 1 & 1 & 0 & 1 \\ 1 & 0 & 2 & 0 \end{bmatrix} \begin{bmatrix} 0 & 1 & 1 & 1 \\ 1 & 0 & 0 & 1 \\ 1 & 1 & 0 & 1 \\ 1 & 0 & 2 & 0 \end{bmatrix} = \begin{bmatrix} 3 & 1 & 2 & 2 \\ 1 & 1 & 3 & 1 \\ 2 & 1 & 3 & 2 \\ 2 & 3 & 1 & 3 \end{bmatrix} = T.$$

The square of the connecting matrix R equals the matrix T for the number of possible two-leg routes among A, B, C, and D.

Similarly,

$$R^3 = R^2 \cdot R = \begin{bmatrix} 3 & 1 & 2 & 2 \\ 1 & 1 & 3 & 1 \\ 2 & 1 & 3 & 2 \\ 2 & 3 & 1 & 3 \end{bmatrix} \begin{bmatrix} 0 & 1 & 1 & 1 \\ 1 & 0 & 0 & 1 \\ 1 & 1 & 0 & 1 \\ 1 & 0 & 2 & 0 \end{bmatrix} = \begin{bmatrix} 5 & 5 & 7 & 6 \\ 5 & 4 & 3 & 5 \\ 6 & 5 & 6 & 6 \\ 7 & 3 & 8 & 6 \end{bmatrix}$$

is a matrix for the number of possible three-leg routes among A, B, C, and D. It can be shown that $R^2 \cdot R = (R \cdot R) \cdot R = R \cdot (R \cdot R) = R \cdot R^2$. In general, the fact that matrix multiplication is associative ensures that multiplication is commutative for powers of a matrix.

The analysis of the graph above provides instances of a theorem in the study of directed graphs: *If R is a matrix for the number of direct (one-leg) routes between the vertices of a directed graph, then R^n is a matrix for the number of n-leg routes between its vertices.* Although a proof of this theorem might not be discussed until a college-level discrete mathematics course, representing graphs by matrices, calculating powers of such matrices, and conjecturing or verifying this theorem are accessible to students in grades 9 to 12 (Peressini et al. 1989; School Mathematics Project 1969, 1970).

Solving Linear Systems

One of the more common uses of matrices is to solve a system of equations. In general, every $n \times n$ linear system with variables $x_1, x_2, \ldots, x_n$ can be represented by a matrix equation of the form

$$MX = C,$$

where M is the coefficient matrix, C is the constant matrix, and X is the $n \times 1$ matrix whose elements are the variables $x_1, x_2, \ldots, x_n$. The matrix equation has a solution

$$X = M^{-1}C$$

if and only if M^{-1}, the multiplicative inverse of M, exists. M^{-1} exists if and only if the determinant of M is not zero.

For instance, the system

$$\begin{cases} 2x - 5y = 18 \\ 3x + y = 10 \end{cases}$$

can be represented as the matrix equation

$$\begin{bmatrix} 2 & -5 \\ 3 & 1 \end{bmatrix} \cdot \begin{bmatrix} x \\ y \end{bmatrix} = \begin{bmatrix} 18 \\ 10 \end{bmatrix}.$$

Left multiplication by the inverse on each side of the matrix equation yields

$$\begin{bmatrix} x \\ y \end{bmatrix} = \begin{bmatrix} \dfrac{1}{17} & \dfrac{5}{17} \\ -\dfrac{3}{17} & \dfrac{2}{17} \end{bmatrix} \begin{bmatrix} 18 \\ 10 \end{bmatrix} = \begin{bmatrix} 4 \\ -2 \end{bmatrix}.$$

Thus, $x = 4$ and $y = -2$ is the solution to the matrix equation and to the original system.

The preceding technique is in theory very powerful. However, because the multiplicative inverse of a matrix larger than 3×3 involves extensive calculations, until recently it has been very difficult to apply in practice.

Fortunately, built-in functions on some calculators can find inverses of matrices as large as 10×10, and computer software can invert much larger matrices (North Carolina School of Science and Mathematics 1988). Hence, technology gives contemporary high school students the power to solve larger systems than college students or professionals might have been able to solve a generation earlier.

Other methods of solving systems, such as Gauss-Jordan row operations on augmented matrices, also exist. However, we prefer to solve systems using the equation-solving technique of this section for two reasons:

- This technique shows more clearly the relationship to solving a simple linear equation in one variable and gives an opportunity to relate procedures in one representation to procedures in an equivalent representation.

- It more clearly focuses on where the properties of matrix multiplication are used and supplies a vehicle for students to understand the logic of algebraic procedures.

Geometry and Transformation Matrices

Matrices provide an additional powerful use in geometry. Any point (x, y) can be represented by the point matrix $\begin{bmatrix} x \\ y \end{bmatrix}$. This idea can be extended to represent any polygon by a $2 \times n$ matrix where n is the number of vertices of the polygon.

Thus, just as matrices store numerical data, in geometry they can be used to represent geometric figures. For instance, the matrix $\begin{bmatrix} 1 & 4 & 1 \\ 1 & 1 & 6 \end{bmatrix}$ represents $\triangle ABC$ in figure 13.7. Note that the matrix $\begin{bmatrix} 1 & 1 & 4 \\ 6 & 1 & 1 \end{bmatrix}$ represents the same triangle but orders the vertices as CAB.

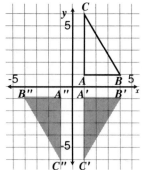

Fig. 13.7

Transformations can also be represented by matrices. Consider the matrix $\begin{bmatrix} 1 & 0 \\ 0 & -1 \end{bmatrix}$. Observe that $\begin{bmatrix} 1 & 0 \\ 0 & -1 \end{bmatrix} \cdot \begin{bmatrix} x \\ y \end{bmatrix} = \begin{bmatrix} x \\ -y \end{bmatrix}$. Hence, the matrix $\begin{bmatrix} 1 & 0 \\ 0 & -1 \end{bmatrix}$

maps the point (x, y) onto $(x, -y)$. Thus, it represents a reflection over the x-axis.

In general, if F is a matrix for some figure and T is a matrix for some transformation of the plane, the product TF, if it exists, represents the image of the figure under that transformation. For instance, if r_x represents a reflection over the x-axis, the statement

$$r_x(\triangle ABC) = \triangle A'B'C'$$

is represented by the matrix equation below:

$$\begin{bmatrix} 1 & 0 \\ 0 & -1 \end{bmatrix} \cdot \begin{bmatrix} 1 & 4 & 1 \\ 1 & 1 & 6 \end{bmatrix} = \begin{bmatrix} 1 & 4 & 1 \\ -1 & -1 & -6 \end{bmatrix}$$

Notice that the image $\triangle A'B'C'$ has vertices $A' = (1, -1)$, $B' = (4, -1)$, and $C' = (1, -6)$, all found in the product matrix.

If each side of the equation above is multiplied on the left by $\begin{bmatrix} -1 & 0 \\ 0 & 1 \end{bmatrix}$,

the result is

$$\begin{bmatrix} -1 & 0 \\ 0 & 1 \end{bmatrix} \cdot \left(\begin{bmatrix} 1 & 0 \\ 0 & -1 \end{bmatrix} \begin{bmatrix} 1 & 4 & 1 \\ 1 & 1 & 6 \end{bmatrix} \right) = \begin{bmatrix} -1 & 0 \\ 0 & 1 \end{bmatrix} \begin{bmatrix} 1 & 4 & 1 \\ -1 & -1 & -6 \end{bmatrix} = \begin{bmatrix} -1 & -4 & -1 \\ -1 & -1 & -6 \end{bmatrix}.$$

The matrix on the far right represents $\triangle A''B''C''$, the reflection image of $\triangle A'B'C'$ over the y-axis. Note that $\triangle A''B''C''$ is also the image of $\triangle ABC$ under a rotation of $180°$ about the origin.

Alternatively, the preceding product could have been calculated using the associative property of matrix multiplication as follows:

$$\left(\begin{bmatrix} -1 & 0 \\ 0 & 1 \end{bmatrix} \cdot \begin{bmatrix} 1 & 0 \\ 0 & -1 \end{bmatrix} \right) \begin{bmatrix} 1 & 4 & 1 \\ 1 & 1 & 6 \end{bmatrix} = \begin{bmatrix} -1 & 0 \\ 0 & -1 \end{bmatrix} \begin{bmatrix} 1 & 4 & 1 \\ 1 & 1 & 6 \end{bmatrix} = \begin{bmatrix} -1 & -4 & -1 \\ -1 & -1 & -6 \end{bmatrix}$$

This instance illustrates an important concept: When matrices for two transformations are multiplied, their product represents the composite of those transformations. Specifically, the preceding example is an instance of the theorem stating that a reflection over the x-axis followed by a reflection over the y-axis is a rotation of $180°$ about the origin. Using composition notation, we can write $r_y \circ r_x = R_{180}$. Here is a geometric context in which to introduce composition of functions, a topic typically studied in a different context in second-year algebra. Again students have an opportunity to use and value the connections among mathematical topics.

The previous discussion raises several questions about the relation between matrices and transformations. For instance, which transformations can be represented by 2 × 2 matrices? What theorems can be derived from a study of matrices for transformations?

In general, if a transformation is linear, that is, if it maps (x, y) onto $(ax + by, cx + dy)$, then it can be represented by the matrix $\begin{bmatrix} a & b \\ c & d \end{bmatrix}$ because

$$\begin{bmatrix} a & b \\ c & d \end{bmatrix} \cdot \begin{bmatrix} x \\ y \end{bmatrix} = \begin{bmatrix} ax + by \\ cx + dy \end{bmatrix}.$$

Many common transformations are linear.

Transformations that preserve distance. The respective matrices for reflection about the x-axis, the y-axis, and the line $y = x$ are shown below:

$$r_x \qquad\qquad r_y \qquad\qquad r_{y=x}$$
$$\begin{bmatrix} 1 & 0 \\ 0 & -1 \end{bmatrix} \qquad \begin{bmatrix} -1 & 0 \\ 0 & 1 \end{bmatrix} \qquad \begin{bmatrix} 0 & 1 \\ 1 & 0 \end{bmatrix}$$

The respective matrices for counterclockwise rotations about the origin of 90°, 180°, 270°, and 360° are as follows:

$$R_{90} \qquad\qquad R_{180} \qquad\qquad R_{270} \qquad\qquad R_{360}$$
$$\begin{bmatrix} 0 & -1 \\ 1 & 0 \end{bmatrix} \qquad \begin{bmatrix} -1 & 0 \\ 0 & -1 \end{bmatrix} \qquad \begin{bmatrix} 0 & 1 \\ -1 & 0 \end{bmatrix} \qquad \begin{bmatrix} 1 & 0 \\ 0 & 1 \end{bmatrix}$$

Transformations that do not preserve distance. The respective matrices for a size change of magnitude a with center at $(0, 0)$ and scale change of a $(a \neq 0)$ units horizontally and b $(b \neq 0)$ units vertically with center $(0, 0)$ are given below:

$$S_a \qquad\qquad S_{a,b}$$
$$\begin{bmatrix} a & 0 \\ 0 & a \end{bmatrix} \qquad \begin{bmatrix} a & 0 \\ 0 & b \end{bmatrix}$$

With trigonometry it is possible to prove that a reflection over any line through $(0, 0)$ can be represented by a 2 × 2 matrix, as can any rotation about the origin as shown in the next section. It is natural to ask if translations or glide reflections (the other two fundamental distance-preserving transformations) can be expressed as 2 × 2 matrices. Surprisingly, they cannot, as we justify below.

Suppose the translation $T_{h,k}$ can be represented by a 2 × 2 matrix $\begin{bmatrix} a & b \\ c & d \end{bmatrix}$. Then the point $(0, 0)$ is mapped to $(0, 0)$, because

$$\begin{bmatrix} a & b \\ c & d \end{bmatrix} \cdot \begin{bmatrix} 0 \\ 0 \end{bmatrix} = \begin{bmatrix} 0 \\ 0 \end{bmatrix}.$$

However, the translation $T_{h,k}$ maps $(0, 0)$ to $(0 + h, 0 + k)$. So unless $h =$

$k = 0$, the translation $T_{h,k}$ cannot be represented by a 2×2 matrix.

It is possible, however, to translate a polygon of n sides using matrix addition. For instance, suppose the matrix $\begin{bmatrix} 5 & 5 & 5 \\ 2 & 2 & 2 \end{bmatrix}$ is added to the matrix

for triangle ABC at the start of this section. This new matrix, $\begin{bmatrix} 6 & 9 & 6 \\ 3 & 3 & 8 \end{bmatrix}$,

represents the image of $\triangle ABC$ under a translation of 5 units to the right and 2 units up. Indeed, the image of any n-gon under the translation $T_{h,k}$ is obtained by adding to the matrix for the original figure a $2 \times n$ matrix with h for each element of the first row and k for each element of the second row.

Alternatively, one can use 3×3 matrices and matrix multiplication to represent translations of the plane. Specifically, represent the point (x, y)

by the matrix $\begin{bmatrix} x \\ y \\ 1 \end{bmatrix}$ and note that $\begin{bmatrix} 1 & 0 & h \\ 0 & 1 & k \\ 0 & 0 & 1 \end{bmatrix} \cdot \begin{bmatrix} x \\ y \\ 1 \end{bmatrix} = \begin{bmatrix} x + h \\ y + k \\ 1 \end{bmatrix}$.

Thus, because $\begin{bmatrix} 1 & 0 & h \\ 0 & 1 & k \\ 0 & 0 & 1 \end{bmatrix}$ maps (x, y) to $(x + h, y + k)$, it is a suitable

matrix for a translation. This latter representation of translations is used frequently in creating computer graphics.

The geometric uses of matrices presented in this section need not all be presented in a geometry course. In geometry, students might simply study transformations without matrix operations. However, in second-year algebra, for example, matrices might be used to review and extend geometric transformations. They provide an opportunity outside geometry to deduce properties of figures and to identify congruent and similar figures. This is, in fact, the approach taken in materials developed by the University of Chicago School Mathematics Project (Coxford, Usiskin, and Hirschhorn 1991; Senk et al., 1990).

Matrices and Trigonometry

The use of matrices with transformations has a powerful payoff in trigonometry. In the previous section we discussed reflections about the x-axis, y-axis, or line $y = x$ and how composites of such reflections yield rotations of $90°$, $180°$, $270°$, or $360°$ about the origin. A rotation about the origin by *any* magnitude can be represented by a 2×2 matrix.

A matrix for R_θ, a counterclockwise rotation of magnitude θ about $(0, 0)$, can be determined by finding the image, under R_θ, of the points $(1, 0)$ and

(0, 1) on the unit circle. Recall that if the point $A = (1, 0)$ is rotated by magnitude θ about the origin, its image is the point $A' = (\cos \theta, \sin \theta)$. To find the image B' of $B = (0, 1)$ under R_θ, first observe that B is the image of A under a rotation of 90°. So B' will be the image of A' under a rotation of 90°. Applying a rotation of 90° to $A' = (\cos \theta, \sin \theta)$ yields $B' = (-\sin \theta, \cos \theta)$. Hence, the general rotation matrix for a counterclockwise rotation of magnitude θ about the origin is

$$R_\theta = \begin{bmatrix} \cos \theta & -\sin \theta \\ \sin \theta & \cos \theta \end{bmatrix}.$$

See figure 13.8.

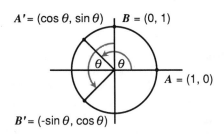

Fig. 13.8

The matrix for R_θ can be very helpful in proving the identities for $\cos(\alpha + \beta)$ and $\sin(\alpha + \beta)$. A rotation of $\alpha + \beta$ can be considered as the composite of a rotation of β followed by a rotation of α. When a rotation of $\alpha + \beta$ is applied to the point $(1, 0)$, the image is $(\cos(\alpha + \beta), \sin(\alpha + \beta))$. A rotation of first β and then α applied to the point $(1, 0)$ gives

$$R_\alpha \circ R_\beta \ (1, 0) = R_\alpha(R_\beta(1, 0))$$
$$= R_\alpha(\cos \beta, \sin \beta).$$

So the matrix for the point $(\cos(\alpha + \beta), \sin(\alpha + \beta))$ is the product of the rotation matrix for α applied to the matrix for the point $(\cos \beta, \sin \beta)$. Thus,

$$\begin{bmatrix} \cos(\alpha + \beta) \\ \sin(\alpha + \beta) \end{bmatrix} = \begin{bmatrix} \cos \alpha & -\sin \alpha \\ \sin \alpha & \cos \alpha \end{bmatrix} \cdot \begin{bmatrix} \cos \beta \\ \sin \beta \end{bmatrix} = \begin{bmatrix} \cos \alpha \cos \beta - \sin \alpha \sin \beta \\ \sin \alpha \cos \beta + \cos \alpha \sin \beta \end{bmatrix}.$$

In order for the matrices to be equal,

$$\cos(\alpha + \beta) = \cos \alpha \cos \beta - \sin \alpha \sin \beta,$$
$$\sin(\alpha + \beta) = \sin \alpha \cos \beta + \cos \alpha \sin \beta,$$

and the desired identities for $\cos(\alpha + \beta)$ and $\sin(\alpha + \beta)$ are obtained.

For students in advanced courses, the use of matrices with trigonometry could be extended to representing vectors as matrices (Schiffer and Bowden 1984) or to representing complex numbers, including the trigonometric form of complex numbers, as matrices (Huber 1988).

CONCLUSION

In this chapter, we have shown that matrices are an exceedingly rich topic and that many uses of matrices are minor extensions or applications of content already in the secondary curriculum. There are several reasons for introducing matrices into the curriculum much earlier than we currently do:

- Matrices introduce a fundamental concept in discrete mathematics.
- They model numerous realistic applications.
- They furnish opportunities for doing arithmetic and algebraic computation in a new context.
- Their properties and their operations lead to important theoretical results.

In addition, matrices are a powerful unifying concept, connecting ideas in mathematics to those in other content domains as well as connecting various branches of mathematics. As such, matrices can be a means of implementing the call of the *Standards* for mathematics to be seen as an integrated whole rather than as a set of isolated topics. Matrices serve as a bridge connecting arithmetic by data storage to graph theory and geometry that, in turn, are another bridge to trigonometry and other advanced topics. If we truly want to show students the beauty of mathematics and its many interrelationships, we can no longer ignore matrices in the secondary school curriculum.

REFERENCES

Coxford, Arthur, Zalman Usiskin, and Daniel Hirschhorn. *Geometry.* Glenview, Ill.: Scott, Foresman & Co., 1991.

Huber, John. "Transformations and Matrices with Applications." In *Projects to Enrich School Mathematics,* edited by Leroy Sachs. Reston, Va.: National Council of Teachers of Mathematics, 1988.

National Council of Teachers of Mathematics. *Curriculum and Evaluation Standards for School Mathematics.* Reston, Va.: The Council, 1989.

North Carolina School of Science and Mathematics, Department of Mathematics and Computer Science. *Matrices.* New Topics for Secondary School Mathematics Series, Materials and Software. Reston, Va.: National Council of Teachers of Mathematics, 1988.

Peressini, Anthony L., Susanna S. Epp, Kathleen A. Hollowell, Susan Brown, Wade Ellis, Jr., John W. McConnell, John Sorteberg, Denisse R. Thompson, Dora Aksoy, Geoffrey D. Birky, Greg A. McRill, and Zalman Usiskin. *Precalculus and Discrete Mathematics.* Field trial ed. Chicago: University of Chicago School Mathematics Project, 1989.

Schiffer, M. M., and Leon Bowden. *The Role of Mathematics in Science.* Washington, D.C.: Mathematical Association of America, 1984.

School Mathematics Project. *SMP, Book C.* London: Cambridge University Press, 1969.

———. *SMP, Book F.* London: Cambridge University Press, 1970.

Senk, Sharon L., Denisse R. Thompson, Steven S. Viktora, Rheta Rubenstein, Judith Halvorson, James Flanders, Natalie Jakucyn, Gerald Pillsbury, and Zalman Usiskin. *Advanced Algebra.* Glenview, Ill.: Scott, Foresman & Co., 1990.

14

A Computer-based Discrete Mathematics Course

Arnold E. Perham
Bernadette H. Perham

S INCE many of the topics in discrete mathematics are closely related to algebra and geometry, they are within the grasp of secondary school students. The data in many of these topics are best represented by arrays. These observations influenced us in selecting and constructing the courseware for a one-semester discrete mathematics course at St. Viator High School in Illinois.

The course has five units: matrix theory, game theory, linear programming, Markov chain theory, and graph theory. Other topics that are sometimes included in discrete mathematics courses, such as combinatorics, trees, probability, and difference equations, were included only to the extent that they support the development and presentation of one of the five major topics.

One of the early evaluators of the course materials, Henry Pollak, remarked that courseware in discrete mathematics should have a degree of open-endedness that allows for the exploration and inclusion of new discoveries. During the course, students have always been encouraged to explore related topics, new algorithms, and alternative methods of displaying and processing information. The students' investigations have led to revisions of the courseware that have kept the units current.

A decade's worth of success in teaching several revisions of the course verifies that the units originally selected are teachable and maintain a high level of interest. Over the years, those who have evaluated and supported the school have been able to relate to the value and goals of the course.

OVERVIEW OF THE CONTENT

An outline of the content covered in each unit can be found in figure 14.1. The capabilities of the supporting software appear in figure 14.2. The discussion gives a sample problem from each of the five units of the course. In each example, the data of the problem are displayed in an array. The manipulation of the data in the arrays, using software programs, depends on the solution sought.

COURSE OUTLINE

I. **Matrix Theory**
   A. Matrix addition, subtraction, multiplication
   B. Scalar multiplication
   C. Determinant of a matrix
   D. Inverse of a matrix
   E. Applications, including the Leontief input-output model for an economy

II. **Game Theory**
   A. Methods for solving strictly determined games
   B. Methods for solving non-strictly determined games
   C. Applications

III. **Linear Programming**
   A. Solution of maximum and minimum problems by graphing
   B. Simplex method
   C. Duality
   D. Applications

IV. **Markov Chains**
   A. Trees
   B. Relation of trees to powers of a matrix
   C. Fixed-point probability vector
   D. Monte Carlo method
   E. Chains with absorptive states
   F. Applications

V. **Graph Theory**
   A. Prim's algorithm for determining a minimal spanning tree
   B. Euler and Hamiltonian circuits
   C. Inferences that can be drawn from the matrix form of a graph
   D. Scheduling as a graph
   E. Shortest path through a graph
   F. Applications

Fig. 14.1

SOFTWARE FOR EACH UNIT

I. **Matrix Theory**

MATRIX ADDITION
   Adds two or more matrices

MATRIX SUBTRACT
   Subtracts two matrices

SCALAR MULTIPLY
   Multiplies a scalar times a matrix

MATRIX MULTIPLY
   Multiplies two or three matrices

DETERMINANT
   Calculates the determinant of a 3×3 matrix

MATRIX INVERSE
   Finds the inverse of a 3×3 matrix

MATRIX INVERSE +
   Finds the inverse of a matrix up to 16×16

II. **Game Theory**

GAMES
  Solves all 2 × 2 games
GAMES BY GRAPHING
  Solves all 2 × 2 games by graphing
EXPECTATIONS
  Calculates expected value for 2 × 2, 2 × 3, and 3 × 2 games
DOMINANCE
  Checks game matrices for dominance
EQUATIONS
  Solves two equations in two variables
EQUATIONS +
  Solves a system of equations in two or more variables

III. **Linear Programming**

GEOMETRY
  Solves LP problems by graphing
SIMPLEX
  Solves LP problems by simplex method
GAMES (see above)
EQUATIONS (see above)

IV. **Markov Chains**

MATRIX POWER
  Raises a matrix to a power
MATRIX MULTIPLY (see above)
MATRIX INVERSE (see above)
RANDOM
  Creates tables of 1000 random numbers
RANDOM WALK
  Solves problems with absorptive states using random numbers
IODA
  Solves problems with absorptive states using a formula

V. **Graph Theory**

SPANTREE
  Calculates the minimal spanning tree
SCHEDULE
  Does critical path analysis
SCHEDULE +
  Alternative algorithm for critical path analysis
REGPOLY +
  Tests if input values will create a regular polyhedron and graphs it
OVERPASSES
  Calculates number of overpasses given number of source and target vertices
MATRIX POWER (see above)

Fig. 14.2

Matrix Theory

Sophisticated questions investigated in advanced courses in matrix theory are not appropriate subject matter for the secondary school level. Nevertheless, there are questions of substance that can be presented to the students. Consider the following example:

A company has a factory in each of two cities: Boise and Butte. Both of these factories employ skilled and unskilled laborers. The data for each of the factories are contained in the following matrix:

$$\begin{array}{cc} & \text{Boise} \quad \text{Butte} \\ \begin{array}{c} \text{skilled} \\ \text{unskilled} \end{array} & \begin{bmatrix} 375 & 523 \\ 529 & 681 \end{bmatrix} \end{array}$$

Given that the average wage of a skilled worker is $712 a week and that for the unskilled worker is $497, what is the total payroll for each plant?

Solution. We can determine the total payroll for each plant by matrix multiplication:

$$[712 \quad 497] \begin{bmatrix} 375 & 523 \\ 529 & 681 \end{bmatrix} = [529913 \quad 710833]$$

The product above shows that the payroll is $529 913 at Boise and $710 833 at Butte.

In practice, the student, using the program MATRIX MULTIPLY, would enter the data for the two matrices to be multiplied. The software gives the student an opportunity to check the data entries and make corrections if necessary and offers the option to have the product matrix reported on the screen alone or on both the screen and the printer.

The technique demonstrated here of using a prematrix multiplier in conjunction with a matrix can be extended to larger matrices in other contexts.

Game Theory

In the previous section the sample problem illustrated how an array of numbers was subjected to the ordinary matrix operation of multiplication. Given another array and another set of questions about the data, a different algorithm might need to be applied for a solution. For example, consider the following problem, which falls under the classification of game theory:

Suppose a town has two pizza places: Joe's and Jane's. Each of them begins an advertising campaign hoping to capture a greater share of the available pizza market. The advertising schemes peculiar to large cities are beyond the means of both Joe and Jane, so they each settle on giving away free pizza slices at the state fair. They both specialize in cheese pizza, with and without pepperoni.

In the matrix below, the rows contain Joe's strategies and the columns contain Jane's strategies. The elements in the matrix represent payoffs for pairs of strategies. These elements are in terms of dollars-a-week increase of sales. A positive value favors Joe, whereas a negative value favors Jane. For example, the entry −$120 means that the strategy of Joe giving away only pizza without pepperoni and Jane giving away only those with pepperoni will increase Jane's sales by $120 a week and Joe, of course, will lose that same amount. How should Joe and Jane play this game?

Jane

		with	without
Joe	with	$100	−$150
	without	−$120	−$200

Solution. First, presume that the data in the matrix are equally available to both owners. It is clear that Joe will never play row 2; that is, he will never give away samples without pepperoni. Jane, in turn, realizing that Joe will never play row 2, will never play column 1. The strategy for this game is for Joe to distribute pizza with pepperoni and Jane to give away pizza samples without pepperoni. The payoff obviously favors Jane; her sales will increase by $150 each week.

This game, as presented, is a strictly determined game, since only one type of pizza will be distributed by each of the parties.

For purposes of illustration, let us change the values in the matrix. The following situation now prevails:

Jane

		with	without
Joe	with	−$50	$100
	without	$150	−$75

In the previous game, since the elements of the first row were greater than the corresponding elements in the second row, Joe chose always to play the first row. Notice that this relationship between row 1 and row 2 does not exist in the new game, nor are the elements in one of the columns greater than or equal to the corresponding elements in the other column. This situation indicates that Joe should employ both strategies, that is, give away pizza with and without pepperoni. But how many of each kind should he give away?

As a means for determining the value of the game and the strategy Joe ought to employ, let us look at some of Joe's options. Presume, for a moment, that Jane stays with the strategy in column 1. Also, presume that 20 percent of Joe's samples are with pepperoni, and the remainder are without. The value of the game can be calculated as follows:

$$0.2(-50) + 0.8(150) = 110$$

Under this strategy, Joe will increase his sales by $110 a week. The following calculations indicate what will happen if Joe's strategy stays the same and Jane switches to column 2:

$$0.2(100) + 0.8(-75) = -40$$

With this strategy, the advantage switches to Jane.

Joe's best strategy is to play in such a way that the value of the game remains the same whether Jane plays column 1 or column 2.

The following array (fig. 14.3) shows calculations similar to those above based on increments of 0.1. The values in the column labeled p indicate the portion of the total number of Joe's pizzas that are *with* pepperoni. It follows that $1 - p$ represents the portion of the total number of Joe's pizzas that are *without* pepperoni. Columns 1 and 2 list the payoffs to Jane or Joe depending on Jane's choice of column 1 or column 2 strategy.

p	$1 - p$	Column 1	Column 2
0	1	+ 150.00	− 75.00
0.1	0.9	+ 130.00	− 57.50
0.2	0.8	+ 110.00	− 40.00
0.3	0.7	+ 90.00	− 22.50
0.4	0.6	+ 70.00	− 5.00
0.5	0.5	+ 50.00	+ 12.50
0.6	0.4	+ 30.00	+ 30.00
0.7	0.3	+ 10.00	+ 47.50
0.8	0.2	− 10.00	+ 65.00
0.9	0.1	− 30.00	+ 82.50
1	0	− 50.00	+ 100.00

Fig. 14.3

From this information, we see that Joe is guaranteed an increase in sales of $30 each week if 60 percent of the samples he distributes are with pepperoni. He is assured this value regardless of the strategy employed by Jane. For Joe to move away from this strategy is risky; the values in the table show that he could do better or he could do worse depending on Jane's choice of strategies.

In practice, the entries in the array above are calculated and displayed for the student by using the program EXPECTATION. In this example, no manipulation of data in the array is required. The student merely interprets the data listed. An alternative method for solving the problem is to use the program GAMES, which uses a different algorithm to find the solution. For more analysis of game theory, see chapter 26.

Linear Programming

Linear programming is a technique useful in discrete optimization. In other words, linear programming specifies the procedure to be followed to minimize or maximize quantities, for example, cost or profit. In the initial approach to the problem, students reduce data to lines that can be plotted on a graph to form a polygon. The coordinates of the vertices of the polygon are then substituted into the expression to be maximized or minimized.

After the basic algorithmic approach to the problem is understood through paper-and-pencil exercises, the student is free to use the program

GEOMETRY, which performs all the calculations and displays the polygonal graph on the screen.

Obviously the above-mentioned technique is not applicable in situations where three or more variables are present. The text and an accompanying program also handle this situation. The program SIMPLEX employs the array concept together with row operations to solve the more complicated problems.

Consider the following scenario: Last year before the Christmas holidays, Sarah and Emory sold board games, giving the proceeds to their local youth club. These board games, all the same type and designed for junior high school students, made a profit of $7 each. The number of boards the students sell is limited only by their time and the amount of material on hand. They have enough material this year to make 150 boards. Last year they sold the 125 they were able to make. Besides selling these boards, they would like to design another, simpler board for younger children, which would take two-thirds the time to make and would sell for a profit of $5. This board, although simpler in design, takes the same amount of material as the original.

How many of each type of board should they make to maximize their profit?

Solution. Let Y represent the number of boards to be made for younger children and J represent the number of boards to be made for junior high school students. The profit equation to be maximized can be written

$$P = 7J + 5Y.$$

The constraints in the problem can be written as a set of inequalities:

Materials constraint: $J + Y \leq 150$

Time constraint: $J + \dfrac{2}{3}Y \leq 125$

$$J \geq 0$$
$$Y \geq 0$$

The inequalities, when graphed in the usual way, form a convex polygonal region (fig. 14.4).

One of the basic premises of linear programming is that the coordinates that maximize the profit equation must necessarily come from one of the vertices of the convex polygon. These coordinates together with the value of the profit equation evaluated with the coordinates in question are listed here:

$A(0, 0)$	Value at A: 0
$B(0, 150)$	Value at B: 750
$C(75, 75)$	Value at C: 900
$D(125, 0)$	Value at D: 875

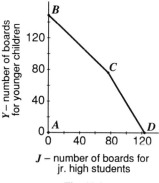

Fig. 14.4

The analysis of the data provided by linear programming suggests that Sarah and Emory should make equal numbers of both board games, that is, 75 of each type. This strategy will yield their highest possible profit, $900.

Markov Chains

Many probability problems can be solved by means of "tree" analysis. The following sample problem shows that the results of the tree analysis can be replicated by applying Markov's algorithm to data arranged in array form:

Sue practices jazz piano after school. She usually practices one or two hours each day. If she plays for an hour on a given day, it is equally likely that she will play one or two hours the next day. However, only one out of five times will she practice for the longer period two days in succession.

The tree in figure 14.5 allows us to answer the following question: Given that it is Monday and Sue practices the piano for one hour, what is the probability that she will practice for two hours on Wednesday?

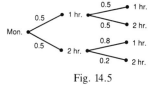

Fig. 14.5

Solution. From the tree, we determine the probability in the usual way:

$$P(2\ h) = 0.25 + 0.10 = 0.35$$

Notice that the same information could be found by using matrix multiplication as shown below. The entries in the first row of the product matrix are the probabilities for practicing one hour and two hours, respectively, on Wednesday evening.

$$
\begin{array}{cc} & \text{1 h} \quad\quad \text{2 h} \end{array} \qquad\qquad\qquad \begin{array}{cc} & \text{1 h} \quad\quad \text{2 h} \end{array}
$$

$$
\begin{array}{c} \text{1 h} \\ \text{2 h} \end{array}
\begin{bmatrix} 0.5 & 0.5 \\ 0.8 & 0.2 \end{bmatrix}
\begin{bmatrix} 0.5 & 0.5 \\ 0.8 & 0.2 \end{bmatrix}
=
\begin{array}{c} \text{1 h} \\ \text{2 h} \end{array}
\begin{bmatrix} 0.65 & 0.35 \\ 0.56 & 0.44 \end{bmatrix}
$$

What happens on Thursday could be calculated by either adding another set of branches to the tree or by raising the original matrix to the third power. The following product matrix contains the results of raising the original to the third power:

$$
\begin{array}{cc} & \text{1 h} \quad\quad \text{2 h} \end{array}
$$

$$
\begin{array}{c} \text{1 h} \\ \text{2 h} \end{array}
\begin{bmatrix} 0.5 & 0.5 \\ 0.8 & 0.2 \end{bmatrix}^{3}
=
\begin{bmatrix} 0.605 & 0.395 \\ 0.632 & 0.368 \end{bmatrix}
$$

The matrix on the right suggests that if it is Monday and Sue practices for one hour, then on Thursday the probability of her practicing one hour or two hours is 0.605 and 0.395, respectively. Notice, too, that row 1 and row 2 of the matrix are beginning to look alike. To see if this phenomenon continues, the original matrix is raised to the tenth power:

$$
\begin{array}{cc} & \text{1 h} \quad\quad \text{2 h} \end{array} \qquad\qquad \begin{array}{cc} & \text{1 h} \quad\quad\quad\quad \text{2 h} \end{array}
$$

$$
\begin{array}{c} \text{1 h} \\ \text{2 h} \end{array}
\begin{bmatrix} 0.5 & 0.5 \\ 0.8 & 0.2 \end{bmatrix}^{10}
=
\begin{array}{c} \text{1 h} \\ \text{2 h} \end{array}
\begin{bmatrix} 0.6153869 & 0.3846131 \\ 0.6153810 & 0.3846190 \end{bmatrix}
$$

The results of the matrix on the right suggest that ten days after Monday, the probabilities of Sue practicing one or two hours are 0.615 and 0.385, respectively, regardless of whether she practices one or two hours on Monday.

Graph Theory

Generally, the data in a graph can be reduced to an array. The array can then be manipulated to uncover information about the graph that often is not obvious. For example, consider the graph in figure 14.6. The information in this graph can be reduced to the matrix shown in figure 14.7. The elements

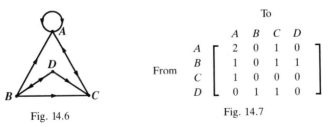

Fig. 14.6   Fig. 14.7

in the matrix indicate the number of ways to go directly from one vertex to another vertex without encountering intervening vertices. Note that A is a loop vertex and that it is possible to move directly from A to A in two different ways, that is, clockwise or counterclockwise.

If the matrix in figure 14.7 above is squared, then the elements will indicate the number of ways to proceed from one vertex to another vertex through one intervening vertex:

To

$$
\text{From}\quad
\begin{array}{c}
 \\ A \\ B \\ C \\ D
\end{array}
\begin{array}{cccc}
A & B & C & D \\
\left[\begin{array}{cccc}
2 & 0 & 1 & 0 \\
1 & 0 & 1 & 1 \\
1 & 0 & 0 & 0 \\
0 & 1 & 1 & 0
\end{array}\right]^2
\end{array}
=
\begin{array}{c}
 \\ A \\ B \\ C \\ D
\end{array}
\begin{array}{cccc}
A & B & C & D \\
\left[\begin{array}{cccc}
5 & 0 & 2 & 0 \\
3 & 1 & 2 & 0 \\
2 & 0 & 1 & 0 \\
2 & 0 & 1 & 1
\end{array}\right]
\end{array}
$$

The matrix on the right indicates that there are three different ways to go from vertex B to vertex A through one intervening vertex. The three ways are these:

1. B to C to A
2. B to A clockwise to A
3. B to A counterclockwise to A

In a similar fashion, the element 0 in row C and column B indicates that it is impossible to go from C to B through an intervening vertex. This information is confirmed by checking it with the diagram in figure 14.6.

This example shows that an airline could use a graph to determine all the ways to go from each city it served to any other city through one intervening city.

FACTORS IN THE CONDUCT OF THE COURSE

The five topics selected for inclusion in a secondary school mathematics course in discrete mathematics determine the prerequisites for a student's enrollment in the course. Essentially, a student should have completed both algebra 1 and geometry and exhibit the same kind of maturity found in the typical algebra 2 student. Since the students receive the necessary software, programming skills are not needed for success in the course. However, substantial access to a computer is required for the student to complete the assignments in the text.

Our course has a computer laboratory with an Apple IIe for each student. In addition, a computer and overhead projector system is available that allows program output to be imaged for classroom viewing.

The basic material in the course is conveyed through lecture, discussion, cooperative learning activities, overhead-projector and computer demonstrations, chalkboard, paper-and-pencil exercises, and the student's work with the computer. The course materials require students to work through the examples with paper and pencil in their first exposure to a topic. After the student has mastered an algorithm, further problems are presented that require the computer programs, selected from a menu, to achieve results that would be almost impossible to obtain in any other way. As a by-product, the student learns both the power and the limitations of computer methods.

Since the computer programs form an integral part of the course, great care was exercised by John Milton, our software editor, to make the programs as user-friendly as possible. Help screens give the students the limits on data and the way in which data must be entered to achieve the desired results.

The evaluation component of the course includes sixteen objective tests. These tests assess the student's knowledge of concepts and applications. Since so much of the course uses the computer, some test items require the student to interpret or make inferences from a computer printout.

IMPLICATIONS FOR OTHERS USING THE MODEL

Successful introduction of a discrete mathematics course into the secondary school curriculum rests on a clear vision of the value of such a course. This requires those charged with the development and implementation of new courses to understand both the relationship of this course to other courses in the mathematics curriculum and what it means to be mathematically literate in the sense proposed in the *Curriculum and Evaluation Standards for School Mathematics* (NCTM 1989). The *Standards* makes clear that to become mathematically literate, students need to experience different mathematics, including concepts and methods of discrete mathematics and the way computers execute mathematical procedures.

The challenge is to integrate the use of the computer into innovative courseware in a meaningful way. We feel that the course described here, a computer-based discrete mathematics course, can serve as a model for those who want to produce course materials that respond to the urgings of the Commission on Standards for School Mathematics and other professional groups, namely, the Conference Board of the Mathematical Sciences (see CBMS [1983]) and the Mathematical Association of America (see Smith [1988]).

BIBLIOGRAPHY

Conference Board of the Mathematical Sciences. *The Mathematical Sciences Curriculum K–12: What Is Still Fundamental and What Is Not.* Report to the NSB Commission on Precollege Education in Mathematics, Science, and Technology. Washington, D.C.: CBMS, 1983.

Hirsch, Christian R., ed. *The Secondary School Mathematics Curriculum.* 1985 Yearbook of the National Council of Teachers of Mathematics. Reston, Va.: The Council, 1985.

National Council of Teachers of Mathematics. *Curriculum and Evaluation Standards for School Mathematics.* Reston, Va.: The Council, 1989.

Smith, David, A., ed. *Computers and Mathematics.* MAA Notes Number 9 of the Mathematical Association of America. Washington, D.C.: MAA, 1988.

15

Using Dominoes to Introduce Combinatorial Reasoning

Jerry Johnson

COUNTING techniques are a fundamental topic in discrete mathematics and have direct applications as general problem-solving tools. Thinking with a combinatorial viewpoint requires students to explore creatively the structural aspects of a problem situation with the hope of reducing it to either a simpler case or a previously solved problem. As a result, many solution possibilities are analyzed systematically, and useful knowledge is gained from both correct and incorrect attempts. The general process is combinatorial reasoning or "the art of counting without counting" (Townsend 1987).

In 1981, a Mathematical Association of America committee published its *Recommendations for a General Mathematical Sciences Program* (MAA 1981). The report's description of a discrete mathematics course emphasized that the process of combinatorial reasoning should be included as an important classroom activity:

> With the right point-of-view, many combinatorial problems have quite simple solutions. However, the object . . . is not to show students simple answers. It is to teach students how to discover such simple answers (as well as not so simple answers). The means for achieving solutions are of more concern than the ends. Learning how to solve problems . . . requires extensive discussion of the logical faults in wrong analyses as much as presenting correct analyses. [Pp. 26–27]

NCTM's *Curriculum and Evaluation Standards for School Mathematics* (1989) suggests that combinatorial reasoning is a useful mathematical tool in a student's collection of representation schemes. However, the reasoning process must involve more than "the application of analytic formulas for permutations and combinations" (NCTM 1989, p. 179).

To achieve the desired results of students being able to use this tool, mathematics teachers must be committed to creating a positive atmosphere that supports combinatorial reasoning. Ample time should be allowed for independent problem solving and group discussion. Also, teachers must structure classroom activities to furnish a conceptual basis for the basic permutation and combination formulas, combinatorial algorithms, and graph theory.

Dominoes and domino games are an excellent resource for problems that encourage combinatorial reasoning. An obvious example is the use of dominoes as a manipulative to illustrate the principle of mathematical induction. However, the world of domino games is much richer and has been relatively unexplored mathematically. Dominoes can be used by almost any age group. If students do not know how to play the basic domino game, they can quickly learn the simple fundamentals.

TYPES OF COMBINATORIAL PROBLEMS

Beckenbach (1964) was perhaps the first to propose a scheme for classifying combinatorial problems. This classification system is a useful guide in the systematic exploration of the structure of a problem. If students can characterize a problem according to its general type, they have made an important step toward its solution. Their attention can then focus on the general solution strategies associated with that type of problem.

To illustrate Beckenbach's classification scheme, consider the problems in organizing a round-robin domino tournament involving 100 contestants and 30 domino "courts."

Existence Problem. Is it possible to arrange the tournament so that no contestant participates in two consecutive games?

Construction Problem. If such a tournament schedule exists, what process can be used to determine it?

Counting Problem. How many such different tournament schedules exist?

Enumeration Problem. How can we systematically list all the possible tournament schedules that meet the desired criteria?

Extremization Problem. What is the most enjoyable tournament schedule, measured by sustained interest through purposely having the best contestants meet each other in the last round?

The numbers originally given—100 contestants and 30 courts—seem irrelevant. A better understanding of the problems and the potential solutions can be gained by reducing the size of the given numbers. Later, the problems and their solutions can be generalized to suit the situation of *n* players and *m* courts.

DOMINO PROBLEM 1

A domino is a rectangle formed by two congruent squares. Each square

Part
6

contains an orderly pattern of "pips" or dots representing a number from zero through six.

Question. Can a domino be made under these constraints?

Answer. This question suggests both an existence and a construction problem, and the answer in each case is an easy yes. Some possible examples are the following:

Question. How many different dominoes can be made under these constraints?

Answer. This question is a counting problem, and the students' attempted solutions to it are often filled with errors. For example, the standard response is forty-nine dominoes, which is the product (7×7) of the number of possibilities for each side of the domino. Other students quickly challenge this answer on the basis that it does not account for duplicates, such as

To resolve the difficulty, the students quickly claim the solution is one-half their previous answer. Then, however, they face the new challenge of explaining how there can be twenty-four and one-half dominoes. Careful combinatorial thinking is now required.

First, two types of dominoes exist: (1) singles, or dominoes whose two square sides bear different numbers, and (2) doubles, or dominoes whose two sides carry the same number. The original counting problem reduces to the two subproblems of determining the number of singles and the number of doubles. Students soon note that their previous logical error can be corrected if they remove the effect of the doubles as follows:

$$7 \text{ doubles} + [(49 - 7)/2] \text{ singles} = 28 \text{ dominoes}$$

A few students will remain uncomfortable with this answer until they have produced the answer to the related enumeration problem.

Question. Can you make a list of the twenty-eight dominoes?

Answer. This question is an enumeration problem, and the students' attempts to produce random lists can be full of gaps and painful to watch. But eventually, a student will develop an organized approach, perhaps placing a six on one side of the domino and systematically listing all possibilities for the other side, blank through six. Then, the students learn to decrease the fixed side by one pip and systematically list all possibilities for the other side except those that duplicate a previous listing. The key is to restrict the nonfixed side to any number less than or equal to the number on the fixed side.

Students have made several observations about the list and its arrangement. Those interested in computer programming have noted the connection of the problem to the following nested counting loop in BASIC:

```
FOR X = 6 TO 0 STEP −1
    FOR Y = 0 TO X STEP 1
        PRINT X, Y
    NEXT Y
NEXT X
```

Other students who counted the number of dominoes associated with each fixed side of a domino saw a connection to the sum

$$7 + 6 + 5 + 4 + 3 + 2 + 1 = 28.$$

DOMINO PROBLEM 2

Suppose four people are going to play a game of dominoes. Under one set of rules, the first step is to divide the full set of dominoes equally among the four players.

Question. How many different ways can this division of the twenty-eight dominoes be made?

Answer. This problem is a counting problem if one agrees that such a division does exist and can be easily performed. Students produce many creative answers to the question. Most of the incorrect solutions that follow hint at some familiarity with factorials and permutation formulas:

- $(a \text{ lot})^4$
- Much more than one million
- $7 \times 6 \times 5 \times 4 = 840$
- $28 \times 27 \times 26 \times 25 = 491\ 400$
- $P(28,4) = 28!/24! = 28 \times 27 \times 26 \times 25 = 491\ 400$
- $P(28,7) = 28!/21! = 28 \times 27 \times \ldots \times 22 = 5.96 \times 10^9$

When asked to explain their answers, students show traces of combinatorial reasoning. Their supportive arguments often reflect either false assumptions (e.g., role of order) or an overly simplistic understanding of the problem.

Two interesting correct solutions are possible, both refinements of the combinatorial logic underlying the incorrect responses above. First, arrange the twenty-eight dominoes in a line, and give each player a sequence of seven dominoes in turn:

          7   7   7   7

Since order is important in the formation of the line, the number of ways of arranging the twenty-eight dominoes is $P(28,28) = 28!/0! = 28!$. However, since each player does not care about the ordering of the seven dominoes received, this effect of $P(7,7) = 7!/0! = 7!$ must be removed four times. Thus, the twenty-eight dominoes can be divided among the four players in

$$28!/[(7!)(7!)(7!)(7!)] = 4.7 \times 10^{14} \text{ ways.}$$

A second approach is to focus on each player at the beginning rather than at the end of the distribution of dominoes. The first player selects seven dominoes randomly out of twenty-eight dominoes. The second player subsequently selects seven dominoes randomly out of the twenty-one remaining dominoes. Similarly, the third player selects seven dominoes randomly out of the fourteen remaining dominoes. The fourth player receives the remaining seven dominoes. Because the product rule is in effect, the total number of different ways of doing this sequential division is $C(28,7) * C(21,7) * C(14,7) * C(7,7)$, or

$$[28!/(21!)(7!)][21!/(14!)(7!)][14!/(7!)(7!)][7!/(0!)(7!)]$$
$$= 28!/[(7!)^4] = 4.7 \times 10^{14}.$$

Some students have a difficult time understanding and accepting the logic behind the two correct solutions because they involve both permutations and combinations. Students object to the second solution because each player does not have the same range of choices. That is, the first player gets to pick seven from twenty-eight, whereas the fourth player picks seven from seven. By their understanding, the given answer should be multiplied by $P(4,4)$ to account for the possible orderings of the four players. One way to overcome these objections is to repeat the problem using a simpler case, such as dividing six dominoes among three players.

Sometimes students want to enumerate systematically the 4.7×10^{14} possible divisions. In his discussion of the problem, Vilenkin (1971) suggests that this desire to use enumeration as a check disappears as soon as students have an intuitive feel for the large size of the numerical answer. If a different division of the twenty-eight dominoes among the four players is made every second, approximately 1.5×10^7 years will be needed to list all the possibilities. A nonexistent (so far) rapid computer and printer capable of listing one million divisions each second would still take about 15 years and miles of paper.

DOMINO PROBLEM 3

After the twenty-eight dominoes are divided among the four players, the task is to play sequentially the complete set of dominoes end to end to form a row. The one restriction is that the touching ends must have the same numerical value.

Question. How many different ways can a set of twenty-eight dominoes be arranged under this restriction? Note that the question essentially asks, How many different games can occur?

Answer. This problem becomes a counting problem if one agrees that it is possible to lay out the full set of dominoes in the prescribed manner. At this point, students need an actual set of dominoes to form their ideas and to test their strategies. Gradually, the students are overwhelmed by the task of arranging twenty-eight dominoes and decide that they must reduce the problem's size if they ever hope to produce a general solution, which clearly involves existence, construction, and counting. They soon learn that the task is not trivial. The new problem is to decide *how* to reduce the problem. It is not sufficient to select a random set of *n* dominoes.

Gardner (1969) suggests a clever solution approach that involves graph theory and this reduction process. First, assume the domino set contains only one domino, such as the double blank:

This set can be arranged in only one way.

Second, assume the domino set is only the three dominoes containing blanks or ones:

By playing with the three dominoes, students easily see that the lineup above is the only one possible if symmetry can be invoked.

Gardner represents this problem by a graph as in figure 15.2. The desired answer is equivalent to the number of ways to trace the graph by a single path without retracing an edge or picking up the pencil. In his graph, each vertex represents a digit (0 or 1) and each edge that connects two vertices represents a domino. Before extending the technique, students need to understand both the graph's representation of the problem and the connection between the tracing of the graph and the total number of domino lineups.

Fig. 15.1. Three-domino graph

Now, assume the domino set is the six dominoes bearing blanks, ones, or twos in figure 15.2. By playing with the six dominoes, students will be able to construct possible lineups, though they will not be confident of finding all the possible lineups. Encourage them to represent the problem with a

Fig. 15.2

graph as in figure 15.3. By tracing this graph, monitoring their decision-making process, and listing possible lineups (see fig. 15.4), students eventually see that little flexibility is involved.

Fig. 15.3. Six-domino graph

[0-0][0-1][1-1][1-2][2-2][2-0]
[1-2][2-2][2-0][0-0][0-1][1-1]
[2-0][0-0][0-1][1-1][1-2][2-2]

Fig. 15.4. Some six-domino lineups

It is as if the dominoes were arranged like a chain around the circumference of a circle, and the primary decision is where to cut the circular chain to open it into a lineup. (See fig. 15.5.) Thus, because six different cuts can be made to determine a starting domino, exactly six different lineups are possible using six dominoes.

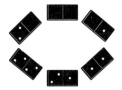

Fig. 15.5

The extension to the ten-domino set involving blanks, ones, twos, and threes is the next step. Again, students should use the dominoes to try to construct a lineup before they draw the related graph. Trying to trace the graph in figure 15.6, students groan as each attempt ends in an error. It is impossible to trace this graph without either retracing an edge or lifting the pencil. As a result, it is impossible to form a lineup using this set of ten dominoes. Because the existence problem is not answered in the affirmative, the problems of construction and counting become irrelevant.

Students should be encouraged to discuss how this graph (fig. 15.6) differs from the others. A major factor is that each of its vertices is of odd degree (5), where the degree of a vertex is the number of edges emanating from it. To help students extend this discovery to the original problem, introduce them to Euler's 1736 solution of the Königsberg bridge problem. In modern terminology, Euler proved that a graph is traceable iff either all its vertices are of even degree *or* exactly two of its vertices are of odd degree.

Fig. 15.6

Given these new insights, students can confront the problem of using a fifteen-domino set involving blanks through fours. By now, students should have learned that the key is first to represent the problem with a graph and then try to trace it (see fig. 15.7). Because each of its vertices is of even degree (6), the graph is traceable and a lineup of the dominoes is possible.

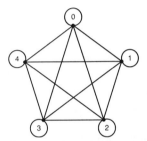

Fig. 15.7. Fifteen-domino graph

The remaining problem is to count the number of possible lineups or paths. In his problem 283, Dudeney (1958) suggests a solution to this problem. Ignoring the use of the doubles, Dudeney claims (unfortunately, without supplying any reasoning) that there are 264 ways to trace the pentagonal portion of the graph. The result for each tracing is a chain, such as

$$[3-4][4-0][0-1][1-2][2-3][3-0][0-2][2-4][4-1][1-3].$$

Each of the five doubles can be inserted in a given chain in two different locations (e.g., [0-0] can be inserted between either [4-0][0-1] or [3-0][0-2]). Also, each chain can be broken at fifteen different locations, since there are fifteen different dominoes that can serve as starting points in the tracing or lineup. Gathering these observations into one equation, we find

(264 chains)(2^5 double-insertions/chain)(15 breaks/chain),

or 126 720, ways to order the fifteen dominoes.

The domino set can now be extended to twenty-one dominoes involving blanks through fives. When drawing the related hexagonal graph, students are quick to point out that each of the six vertices is of odd degree (7).

Thus, by Euler's theorem, the graph cannot be traced, and the twenty-one dominoes cannot be arranged in a line.

The final situation is the original problem using the full set of twenty-eight dominoes. The associated graph is heptagonal in shape and each vertex is of even degree (8). Thus, lineups can be constructed. Because the combinatorial reasoning is quite complex, Gardner merely suggests that the total number of lineups is 7 959 229 931 520. Using Dudeney's reasoning, we find that there are seven doubles to be inserted and that each chain can be broken in twenty-eight locations. The difficult task is to determine the number of ways to trace the heptagonal portion of the graph. Because no algorithm exists for this tracing, the author and two graph theorists are trying to replicate Gardner's results using techniques other than case-by-case exhaustion. Perhaps you and your students would like to pick up this challenge as well.

SOME FINAL SUGGESTIONS

These domino problems and their analyses are but a beginning for students of combinatorial reasoning. Armanino (1977) offers a large collection of domino games to explore. Other interesting domino problems can be found in Vilenkin (1971), Gardner (1969), and Dudeney (1958). Finally, variations on the domino game can be tried, such as Triominoes and other related board games.

By exploring the domino problems in this chapter, students begin to recognize situations that require combinatorial reasoning. The students also gain an appreciation for the power of the mathematical tools that support combinatorial reasoning, such as permutations, combinations, graph theory, algorithms, and general problem-solving techniques. In sum, the students are introduced properly to the art of counting without counting.

REFERENCES

Armanino, Dominic. *Dominoes: Games, Rules, & Strategy.* New York: Simon & Schuster, 1977.

Beckenbach, Edwin. *Applied Combinatorial Mathematics.* New York: John Wiley & Sons, 1964.

Dudeney, H. E. *Amusements in Mathematics.* New York: Dover Publications, 1958.

Gardner, Martin. "A Handful of Combinatorial Problems Based on Dominoes." *Scientific American,* (December 1969), pp. 122–27.

Mathematical Association of America, Committee on the Undergraduate Program in Mathematics. *Recommendations for a General Mathematical Sciences Program.* Washington, D.C.: MAA, 1981.

National Council of Teachers of Mathematics. *Curriculum and Evaluation Standards for School Mathematics.* Reston, Va.: The Council, 1989.

Townsend, Michael. *Discrete Mathematics: Applied Combinatorics and Graph Theory.* Menlo Park, Calif.: Benjamin-Cummings Publishing Co., 1987.

Vilenkin, N. Ya. *Combinatorics.* New York: Academic Press, 1971.

16

Combinatorics and Geometry

Vincent P. Schielack, Jr.

S TUDENTS often face an overwhelming difficulty when studying combinatorics. They are given many formulas (with little justification), they
attempt to memorize these formulas (with no relational framework), and
the resulting confusion can be disastrous. Since most of these students are
already relatively comfortable with elementary geometry, geometric topics
can be used as a starting point to motivate and develop students' combinatorial sense rather than rely on rote memorization. This method makes the
elementary geometry course a natural place in the secondary school mathematics curriculum to integrate counting procedures.

Two especially fruitful areas where geometry and combinatorics interface
are those dealing with elementary permutations and combinations and those
regarding sums of positive integers. For the latter, area or volume formulas
can be used to establish formulas such as

$$1 + 2 + \cdots + n = \frac{n(n + 1)}{2}$$

and

$$1^2 + 2^2 + \cdots + n^2 = \frac{n(n + 1)(2n + 1)}{6}.$$

Since this presentation is available elsewhere (Schielack 1987), we shall
concentrate on permutations and combinations, using a series of geometric
questions to derive combinatorial results. These counting results are necessary to answer *geometric* questions, so that the geometry and combinatorics form a somewhat symbiotic relationship—elementary geometric ideas
are used to develop combinatorial procedures that in turn yield solutions to
more advanced geometric questions. Validation is an important issue, too.
However, the focus of this chapter is on deriving the results.

Please keep in mind that the questions represent a problem-solving experience for students. The questions should be presented with some general
techniques that can lead toward solutions. Rather than tackling a difficult
problem in all its generality, it is often helpful to try some special cases with

simple solutions until a conjecture can be formulated and tested. Another facilitating procedure is to reduce the problem to one or more simpler problems that have already been solved.

Question 1. Given n points, how many distinct *names* of line segments are there with two of the n points as endpoints? (Using the standard notation $\overline{AB}$ for the line segment with endpoints A and B, we note that there are two names for this segment, $\overline{AB}$ and $\overline{BA}$.)

It is easy and instructive to try some special cases. (Note that it is immaterial how the points are configured.) For example, if $n = 4$ and the points are named A, B, C, and D, here are all possibilities:

$\overline{AB}$	$\overline{BA}$	$\overline{CA}$	$\overline{DA}$
$\overline{AC}$	$\overline{BC}$	$\overline{CB}$	$\overline{DB}$
$\overline{AD}$	$\overline{BD}$	$\overline{CD}$	$\overline{DC}$

Any of the four letters can be the first in a name, and any of the remaining three can be the second, yielding a total of 4·3, or 12, different names of segments. It is now a small step to reach the answer to our original question:

$$n(n - 1) \text{ distinct names}$$

Question 2. Given n points, no three of which are collinear, how many distinct *names* of triangles are there having three of the n points as vertices? (Using the standard notation, we realize that $\triangle ABC$ and $\triangle BCA$ are different names for the same triangle.)

Let's list the possibilities for the four points A, B, C, and D.

$\triangle ABC$	$\triangle BAC$	$\triangle CAB$	$\triangle DAB$
$\triangle ABD$	$\triangle BAD$	$\triangle CAD$	$\triangle DAC$
$\triangle ACB$	$\triangle BCA$	$\triangle CBA$	$\triangle DBA$
$\triangle ACD$	$\triangle BCD$	$\triangle CBD$	$\triangle DBC$
$\triangle ADB$	$\triangle BDA$	$\triangle CDA$	$\triangle DCA$
$\triangle ADC$	$\triangle BDC$	$\triangle CDB$	$\triangle DCB$

Here, we note that any of the four letters can be the first in a name; then any of the remaining three can be the second; and finally, either of the last two can be the third letter in the name. This results in a total of 4·3·2, or 24, different names of triangles. The answer for n points follows easily:

$$n(n - 1)(n - 2)$$

The next question can be used to stimulate an answer to the general permutation problem, that is, find $P(n,m)$, the number of permutations of n objects taken m at a time, where $m \leq n$.

Question 3. Given a set S of n points, no three of which are collinear, find the number of different one-directional paths of length m in S. (A *path* of length m, where $m \leq n$, begins at one of the n points, proceeds through $m - 2$ intermediate points in S, and ends at another point of S, all without duplicating any of the points used.)

Questions 1 and 2 are special cases of this question; the name of a line segment is equivalent to a path of length 2, and the name of a triangle is equivalent to a path of length 3. Let's use the particular case $n = 10$, $m = 4$. There are ten choices for the starting point. We then have nine choices for the second point (since we can't duplicate points), eight choices for the third point, and seven choices for the ending point. Thus, there are a total of $10 \cdot 9 \cdot 8 \cdot 7$ paths of length 4.

The results of a few more examples will lead to the conclusion that

$$n(n - 1)(n - 2)\cdots(n - (m - 1)) = \frac{n!}{(n - m)!} = P(n,m)$$

different one-directional paths of length m exist in a set of n points. (Here is the natural place to define $n!$, since this represents the number of one-directional paths of length n.) However, the formula for $P(n,m)$ is for illustrative purposes only; we do not advocate its memorization. You might use this closed formula in a computer program, but from an operational standpoint, it is better to derive the formula on an ad hoc basis each time it is needed rather than to commit it to memory. This method also imparts a "feel" for the situation in a particular problem.

Question 4. Given a set of n points, how many distinct line segments are there with two of the n points as endpoints?

This is equivalent to asking how many distinct sets of two points there are, since two points determine a unique line segment. Let's take $n = 7$ as a special case and call the seven points A, B, C, D, E, F, and G. First, we can list all segments with point A as an endpoint—$\overline{AB}$, $\overline{AC}$, $\overline{AD}$, $\overline{AE}$, $\overline{AF}$, and $\overline{AG}$. Next, we list all segments containing B—$\overline{BA}$, $\overline{BC}$, $\overline{BD}$, $\overline{BE}$, $\overline{BF}$, and $\overline{BG}$. In this way, we list six segments for each of the seven possible "pivot" points, making a total of $7 \cdot 6$, or 42, names of line segments. However, each segment has two different names (e.g., $\overline{AB}$ and $\overline{BA}$ name the same segment), and thus appears twice on the list of forty-two. So we may obtain the number of distinct segments by dividing 42 by our overcounting factor of 2, giving us an answer of 21. Another example or two will indicate the answer to the n-point question, $\dfrac{n(n - 1)}{2}$ segments, reasoned thus: there are n choices for the first point in the name of a line segment, and $n - 1$ choices for the second point. Since each line segment is named twice, we divide by 2 to obtain the result.

Here is an alternative method for finding this answer. Enumerate the points 1 through n. There are $n - 1$ segments having the first point as an endpoint. There are $n - 2$ segments containing the second point but not the first. There are $n - 3$ segments with the third point but neither of the first two. . . . Finally, one segment contains the final two points as endpoints. Therefore, our number of segments is $1 + 2 + \cdots + (n - 1)$, which equals $\dfrac{n(n - 1)}{2}$ by the formula for the sum of the first $n - 1$ positive integers. Therefore, we have a connection between this sum and a certain combination:

$$1 + 2 + \cdots + (n - 1) = \frac{n(n - 1)}{2} = C(n,2),$$

the number of combinations of n objects taken two at a time.

Question 5. Given a set of n points, no three of which are collinear, how many distinct triangles are there having three of the n points as vertices?

Equivalently, how many distinct sets of three points are there? Let us proceed from our answer to question 4, using $n = 7$ again. To each of the twenty-one segments found in question 4, we add one of the remaining five points (those other than the two endpoints of the segments) to determine a triangle. One difficulty remains: the same triangle is determined by using, for example, $\overline{AB}$ and point C, $\overline{AC}$ and point B, or $\overline{BC}$ and point A. In fact, each triangle is being counted three times. We can correct this by dividing by our overcounting factor of 3. Thus, the number of triangles is the number of segments, $\dfrac{7 \cdot 6}{2}$, times the number of remaining points, 5, divided by our overcounting factor, 3. This yields $\dfrac{7 \cdot 6 \cdot 5}{2 \cdot 3} = 35$. For n points, the result will be $\dfrac{n(n - 1)(n - 2)}{2 \cdot 3} = C(n,3)$.

This result can also be obtained by using the answer to question 2. There are $n(n - 1)(n - 2)$ distinct names of triangles with vertices from the given set of n points. A particular set of three of these points determines a unique triangle, but this triangle has 3! different names, one for each ordering of the three vertices. So we have overcounted by a factor of 3!, and the number of different triangles is therefore $\dfrac{n(n - 1)(n - 2)}{3!}$.

Question 6. Given a set of n points, no four of which are coplanar, how many distinct tetrahedra are there having four of the n points as vertices?

Equivalently, we need to find the number of distinct sets of four points. We can use our answer to question 5. When $n = 7$, we can start with one of the thirty-five triangles and add one of the four remaining points (those other than the triangle's vertices) to determine a tetrahedron. However, 35·4, or 140, is too large by a factor of 4, since each tetrahedron was counted four times. (For example, the tetrahedron with vertices A, B, C, and D results from $\triangle ABC$ and point D, $\triangle ACD$ with point B, $\triangle ABD$ with point C, or $\triangle BCD$ with point A.) Thus, the number of tetrahedra is the number of triangles, $\dfrac{7 \cdot 6 \cdot 5}{2 \cdot 3}$, times the number of remaining points, 4, divided by the overcounting factor 4; the result is $\dfrac{7 \cdot 6 \cdot 5 \cdot 4}{2 \cdot 3 \cdot 4} = 35$. For n points, the result is $\dfrac{n(n-1)(n-2)(n-3)}{2 \cdot 3 \cdot 4} = C(n, 4)$. To obtain the general formula for $C(n,m)$, the number of combinations of n objects taken m at a time, where $m \le n$, we merely continue the pattern established by answering these questions. Unfortunately, our physical dimensions are exhausted, but the same effect can be elicited by considering question 7, which could also have been used in place of questions 4–6.

Question 7. Given set S of n points on a circle, how many convex m-gons are there with vertices contained in S?

This is equivalent to asking the number of different sets of m points that can be chosen from the n points in S (or, in the terminology of combinations, the value of $C(n,m)$). Questions 4, 5, and 6 essentially answered this question for $m = 2$, 3, and 4, respectively; the answers were $\dfrac{n(n-1)}{2}$, $\dfrac{n(n-1)(n-2)}{2 \cdot 3}$, and $\dfrac{n(n-1)(n-2)(n-3)}{2 \cdot 3 \cdot 4}$.

For the particular case $n = 7$ and $m = 5$, we obtain

$$C(7,5) = \frac{7 \cdot 6 \cdot 5 \cdot 4 \cdot 3}{2 \cdot 3 \cdot 4 \cdot 5}$$

from question 6 and the pattern we have already established. This could also be written as $C(7,5) = \dfrac{7 \cdot 6}{2} \left(\dfrac{5 \cdot 4 \cdot 3}{3 \cdot 4 \cdot 5} \right) = C(7,2)$.

This clearly illustrates the principle $C(n,m) = C(n, n - m)$. (Our specific examples in questions 5 and 6 show $C(7,3) = C(7,4) = 35$.)

The general formula

$$C(n,m) = \frac{n!}{m!(n-m)!} = \frac{P(n,m)}{m!}$$

can be readily established using our pattern and a little algebra involving factorials. The formula points to what we have found in exploring questions

4–7: The number of ways of choosing m objects from a set of n objects, disregarding order, can be found by counting the number of ways of selecting a permutation of m objects and dividing by our overcounting factor of $m!$, the number of ways the m objects can be ordered.

Some additional questions using geometry and combinatorics follow. Solutions appear in brackets.

Question 8. How many points of intersection are formed by n coplanar lines if no two lines are parallel and no three intersect in a common point? $\left[\dfrac{n(n-1)}{2}, \text{ or } C(n,2) \right]$ Into how many regions do these lines divide the plane? $\left[\dfrac{n(n+1)}{2} + 1 \right]$ How many of these regions are finite? $\left[\dfrac{(n-1)(n-2)}{2} \right]$ How many are infinite? $[2n]$

Question 9. How many diagonals does a convex n-gon have? $\left[\dfrac{n(n-3)}{2} \right]$ How many triangles are determined by the vertices of a convex n-gon? $[C(n,3)]$ How many of these triangles do not have a side coinciding with a side of the polygon? $\left[\dfrac{n(n-4)(n-5)}{6}, \text{ or } C(n,3) - n(n-4) - n, \text{ for } n \geq 4 \right]$

Litwiller and Duncan (1987) offer more geometric counting.

BIBLIOGRAPHY

Finkbeiner, Daniel T. II, and Wendell D. Lindstrom. *A Primer of Discrete Mathematics.* New York: W. H. Freeman & Co., 1987.

Graham, Ronald L., Donald E. Knuth, and Oren Patashnik. *Concrete Mathematics: A Foundation for Computer Science.* Reading, Mass.: Addison-Wesley Publishing Co., 1989.

Grimaldi, Ralph P. *Discrete and Combinatorial Mathematics: An Applied Introduction.* Reading, Mass.: Addison-Wesley Publishing Co., 1989.

Litwiller, Bonnie H., and David R. Duncan. "Geometric Counting Problems." In *Learning and Teaching Geometry, K–12,* 1987 Yearbook of the National Council of Teachers of Mathematics, edited by Mary M. Lindquist. Reston, Va.: The Council, 1987.

Page, E. S., and L. B. Wilson. *An Introduction to Computational Combinatorics.* London: Cambridge University Press, 1979.

Schielack, Vincent P., Jr. "Mathematical Applications of Geometry." In *Learning and Teaching Geometry, K–12,* 1987 Yearbook of the National Council of Teachers of Mathematics, edited by Mary M. Lindquist. Reston, Va.: The Council, 1987.

17

Counting
with Generating Functions

Lisa J. Evered
Brian Schroeder

G ENERATING functions were discovered and named by the great French mathematician Pierre-Simon de LaPlace. His *Théorie analytique des probabilités* (1812) included the first systematic treatment of generating functions applied to probability.

The concept of a generating function is a simple one. However, before computers were readily available, the algebraic manipulations associated with generating functions were too difficult for secondary school students. In a computer environment many of these difficulties can be avoided.

The purpose of this chapter is to present ordinary and exponential generating functions as alternatives to classic combinatorial formulas. Another purpose is to suggest ways in which students' computer competencies can be combined with the method of generating functions to solve nontrivial counting problems.

A generating function is an algebraic expression. The coefficients of the terms of the expression give solutions to a sequence of related counting problems. In a polynomial, or ordinary, generating function,

$$g(x) = a_0 + a_1 x + a_2 x^2 + \ldots + a_r x^r + \ldots,$$

the coefficients $a_0, a_1, a_2, \ldots, a_r$ are answers to the problems under investigation.

Perhaps the most familiar of all generating functions is the binomial formula

$$(1 + x)^n = C(n,0) + C(n,1)x + C(n,2)x^2 + \ldots + C(n,n)x^n.$$

Each of the coefficients, $C(n,0), C(n,1), C(n,2), \ldots, C(n,n)$, is an answer to a counting problem, namely, the number of ways to choose r objects from n objects. To see the relationship between the expansion of a binomial and this family of counting problems, consider a specific situation, say $(1 + x)^3$.

Rewrite the factors in the product $(1 + x)^3 = (1 + x)(1 + x)(1 + x)$ vertically as follows:

$$\begin{Bmatrix} 1 \\ x \end{Bmatrix} \quad \begin{Bmatrix} 1 \\ x \end{Bmatrix} \quad \begin{Bmatrix} 1 \\ x \end{Bmatrix}$$

The coefficient of the x^2 term in the expansion is precisely the number of ways possible to pass through these factors selecting exactly two x's from the three x's present. The three ways in figure 17.1 are the only ways to pass through the factors selecting exactly two x's; hence, the coefficient of the x^2 term in the expansion must be 3.

START $\cdots\begin{Bmatrix} 1 \\ x \end{Bmatrix}\cdots\begin{Bmatrix} 1 \\ x \end{Bmatrix}\cdots\begin{Bmatrix} 1 \\ x \end{Bmatrix}\cdots>$ END

START $\cdots\begin{Bmatrix} 1 \\ x \end{Bmatrix}\cdots\begin{Bmatrix} 1 \\ x \end{Bmatrix}\cdots\begin{Bmatrix} 1 \\ x \end{Bmatrix}\cdots>$ END

START $\cdots\begin{Bmatrix} 1 \\ x \end{Bmatrix}\cdots\begin{Bmatrix} 1 \\ x \end{Bmatrix}\cdots\begin{Bmatrix} 1 \\ x \end{Bmatrix}\cdots>$ END

Fig. 17.1

It would be tedious to trace the factors for *each* individual counting problem where r objects are to be selected from n objects. However, once the generating function is expanded using paper-and-pencil algebra or a computer program, the coefficients found through this algorithmic process furnish answers for the entire class of problems "n choose r." Using the same analysis of factors, we can devise a generating function for counting problems that are not as routine, as the following examples illustrate:

Example 1. *USA Today* wants to publish six interviews about a Supreme Court decision. The twenty people interviewed were six white females, three black females, seven white males, and four black males. The article must include an interview from at least one person in each category. For example, one way of selecting the six interviews is two white females, two black females, one white male, and one black male. How many ways can the newspaper select the six interviews for publication?

The task is to write a generating function whose coefficients are the answers not only to the problem of selecting six interviews but also to the problem of choosing 4, 5, 6, 7, . . . , or 20 interviews for publication. Begin by writing a vertical factor for each category of interviewees. The factor for the first category, white females, is shown on the far left in figure 17.2.

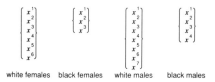

white females   black females   white males   black males

Fig. 17.2

Six white females are interviewed, and the newspaper must include at least one in the published interviews. The exponents of the terms of this factor inventory the possibilities—one white female, two white females, . . . , six white females—for this category. Using similar reasoning, we list the vertical factors for all the categories, as in figure 17.2.

The solution to the six-interview question is the number of ways of passing through all four vertical factors to obtain an exponent sum of 6 (see fig. 17.3).

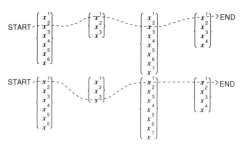

Fig. 17.3

Tracing all possible paths through these vertical factors to obtain an exponent sum of 6 would indeed be tedious. If *USA Today* wanted to publish ten interviews, finding all possible paths through the factors with an exponent sum of 10 would be even more tiring. It is relatively simple, however, to multiply the factors

$$(x^1 + x^2 + x^3 + x^4 + x^5 + x^6)(x^1 + x^2 + x^3)$$
$$\cdot (x^1 + x^2 + x^3 + x^4 + x^5 + x^6 + x^7)(x^1 + x^2 + x^3 + x^4)$$

to obtain the generating function

$$x^4 + 4x^5 + 10x^6 + 19x^7 + 30x^8 + 42x^9 + 53x^{10} + 61x^{11} + 64x^{12}$$
$$+ 61x^{13} + 53x^{14} + 42x^{15} + 30x^{16} + 19x^{17} + 10x^{18} + 4x^{19} + x^{20}.$$

The solution to the six-interview problem is the coefficient, 10, of the term x^6 in the generating function. Your computer-capable students may enjoy creating a Pascal procedure or a BASIC or "C" program to perform the polynomial multiplication.

Example 2. In a fifteen-round boxing match, three judges each award 1 point to the fighter they believe to have won the round. If no rounds are scored as draws, how many ways can the champion receive exactly 29 points? One possibility is 15 from the first judge, 10 from the second judge, and 4 from the third judge. To win the fight, at least two judges must score 8 or more points for the winner.

To write an ordinary generating function for this problem, construct a polynomial factor for each scorer with exponents that inventory the number of rounds that could be awarded to the champion. Since there are three judges with identical scoring possibilities, the required factor for each judge is

$$(1 + x^1 + x^2 + \ldots + x^{15}).$$

The number of ways the champ can receive exactly 29 points is the coefficient, 150, of the term x^{29} in the expansion of

$$(1 + x^1 + x^2 + \ldots + x^{15})^3.$$

If the champ wins 29 of 45 possible points, is it still possible for him to lose the fight? [Yes. For example, the first two judges could each award 7 rounds to the champ and 8 rounds to the challenger, while the third judge could award 15 rounds to the champ and 0 rounds to the challenger. In all, the champ would receive 29 points, but lose the fight on a 2 to 1 split decision.] How many ways can the champ lose the fight when scoring 29 points? [A 7-7-15 scoring for the champ by the three judges is the only situation where the champ could receive 29 points and still lose the fight. Since there are three judges who could award all 15 rounds to the champ, there are three ways that the champ could lose when receiving 29 points.]

The generating functions for the interview question and the boxing question are *polynomial*, or *ordinary*, generating functions. Ordinary generating functions are used in problems where order is not important. To find a similar tool to solve problems where order is critical, recall the following relation between the number of combinations of n things taken r at a time, $C(n,r)$, and the number of permutations of r things selected from n, $P(n,r)$:

$$C(n,r) = \frac{n!}{r! \, (n - r)!} = \frac{1}{r!} P(n,r)$$

Return to the binomial formula

$$(1 + x)^n = C(n,0) + C(n,1)x + C(n,2)x^2 + \ldots + C(n,n)x^n$$

and replace $C(n,r)$ with $1/r!P(n,r)$ for $0 \le r \le n$ to obtain

$$(1 + x)^n = P(n,0) + P(n,1)\frac{x}{1!} + P(n,2)\frac{x^2}{2!} + \ldots + P(n,n)\frac{x^n}{n!}.$$

Because of the similarity between this sequence and the sequence

$$e^x = 1 + \frac{x}{1!} + \frac{x^2}{2!} + \ldots ,$$

the altered version of the binomial theorem is called an *exponential* generating function.

The coefficients of this function count the permutations rather than the combinations of r objects chosen from n. Note that the coefficient required for, say, the rth term is $P(n,r)$. The $r!$ remains attached to the variable x^r and is not taken as part of the coefficient. The next examples illustrate the use of an exponential generating function to solve counting problems where order *is* important.

Example 3. RNA, the building block of heredity, is a chain consisting of links. Each link is one of four possible bases: adenine (A), cytosine (C), guanine (G), or uracil (U). The order in which each base is linked is important. How many possible RNA chains of length 4 are there if two A's, two G's, three C's, and one U are available? Following the reasoning used in the previous example, write a factor for each base.

$$\left(1 + \frac{x}{1!} + \frac{x^2}{2!}\right) \qquad \left(1 + \frac{x}{1!} + \frac{x^2}{2!}\right) \qquad \left(1 + \frac{x}{1!} + \frac{x^2}{2!} + \frac{x^3}{3!}\right) \qquad \left(1 + \frac{x}{1!}\right)$$

$$\text{Base A} \qquad\qquad \text{Base G} \qquad\qquad\qquad \text{Base C} \qquad\qquad\quad \text{Base U}$$

Note the inclusion of the factorial denominators that characterize an exponential generating function. Again, the exponents of each factor list the possibilities for each base, A, G, C, or U. The number of RNA chains of length 4 is precisely the coefficient, 162, of the term $x^4/4!$ in the product.

To compute this coefficient, your students may wish to develop a Pascal program for exponential generating functions.

Example 4. Companies in California, Texas, and New York are competing for ten separate defense contracts. Their senators have agreed that each state should receive at least one contract and no more than five. How many ways can the contracts be awarded to satisfy the senators' restrictions? Consider a sequence of contracts each marked with the name of the state in which it will be awarded (see fig. 17.4).

Fig. 17.4

Order is important because the two groups represent different assignments. Hence, the generating function needed must be exponential. Since each state receives at least one contract and not more than five, the required generating function is

$$\left(\frac{x}{1!} + \frac{x^2}{2!} + \ldots + \frac{x^5}{5!}\right)^3.$$

The answer to the ten-contract problem is exactly the coefficient, 44 730, of the term $\dfrac{x^{10}}{10!}$ in the product.

CONCLUSION

The method of ordinary and exponential generating functions is a powerful and convenient tool for solving a variety of combinatorial problems. It gives the teacher an opportunity to integrate traditional algebra content, new discrete mathematics content, and computer use. Later, as students study probability and statistics, their experience with generating functions should be an excellent foundation for the understanding of moment generating functions and their importance in finding solutions to families of probabilistic problems.

BIBLIOGRAPHY

Barnier, William, and Jean Chan. *Discrete Mathematics with Applications.* New York: West Publishing Co., 1989.

Beckenbach, Edwin. *Applied Combinatorial Mathematics.* New York: John Wiley & Sons, 1964.

Grimaldi, Ralph P. *Discrete and Combinatorial Mathematics.* New York: Addison-Wesley Publishing Co., 1989.

LaPlace, Pierre-Simon. *Théorie analytique des probabilités.* Paris: Mme ve. Courcier, 1812.

Riordan, John. *An Introduction to Combinatorial Analysis.* New York: John Wiley & Sons, 1958.

Roberts, Fred S. *Applied Combinatorics.* Englewood Cliffs, N.J.: Prentice-Hall, 1984.

Tucker, Alan. *Applied Combinatorics.* New York: John Wiley & Sons, 1984.

18

Incorporating Recursion and Functions in the Secondary School Mathematics Curriculum

Robert H. Cornell
Edward Siegfried

T HERE are at least two ways in which the development of computing affects the mathematics teaching profession. First, the contemporary practice of mathematics by researchers, technicians, users of statistics, and others has come to depend on computers. Second, the understanding of mathematics can be supported by computers, calculators, and other technologies. The NCTM *Curriculum and Evaluation Standards for School Mathematics* (1989) reflects these twin influences. The *Standards* specifies the changes in curricular emphasis needed to prepare students better for the workplace they are likely to find, and it calls for the use of technological tools at all levels of instruction.

We hope to contribute to the implementation of the *Standards* by showing how ideas from discrete mathematics can offer a new view of traditional content, and how this content can be made accessible using available computer-based tools. Our discussion uses the most familiar software—Logo and spreadsheets—and is hardware independent. We begin with sequences, a topic recommended for increased emphasis in standard 12 on discrete mathematics for grades 9 through 12.

Ask a beginning algebra student to describe the sequence 8, 11, 14, 17, 20, 23, 26, . . . and the response will most likely be, "You take any term and add 3 to it to get the next one." Although the statement is true, it does not specify this sequence adequately, since what was said is true of other sequences as well. This same algebra student will easily agree that a *complete* specification of the sequence is, "The first term is 8, and each term after the first is 3 more than the term before it." This recursive description is more natural than the more commonly used closed form "$3n + 5$" where n is the term position.

Arithmetic sequences such as the one above are the best initial examples

149

to show how recursive expressions can be modeled using Logo or spreadsheets. If we let T denote that sequence, we may express the function T easily in Logo:

```
TO T :N
    IF :N = 1 [ OUTPUT 8 ]
    IF :N > 1 [ OUTPUT 3 + T(:N − 1) ]
END
```

To create a spreadsheet for a recursively defined sequence, it is helpful to have the index numbers 1, 2, 3, . . . in column A and the values of the terms of the sequence in column B. For the sequence T, enter the number 8 in cell B1 and the formula 3 + B1 in cell B2. Replicate this formula down column B to obtain figure 18.1.

	A	B	C	D
1	1	8		
2	2	11		
3	3	14		
4	4	17		
5	5	20		
6	6	23		
7	7	26		
8	8	29		
9	9	32		
10	10	25		
11	11	38		

Fig. 18.1

Both of these models of the sequence mimic the conventional recursive definition of the function T:

$$T(n) = \begin{cases} 8 & \text{if } n = 1 \\ 3 + T(n − 1) & \text{if } n > 1 \end{cases}$$

RECURSIVELY DESCRIBED SITUATIONS

The examples in this section are suitable for deepening students' understanding of functions and recursion. Our discussion indicates the outcome of student investigation of each situation, not the details of the process. The students need time to explore, form conjectures, test the conjectures, record their discoveries, and present their ideas to others.

The Tower of Hanoi

The problem

Given three posts and disks of varying diameters, how many moves are needed to transfer n disks from one post to another? (See fig. 18.2.) A "move" is shifting a single disk from one post to another, under the restriction that no disk may be placed on a disk of smaller diameter. "How many" means the least number of moves generated by any algorithm for making the transfer.

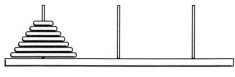

Fig. 18.2

Modeling the problem with a Logo function

Let $H(n)$ denote the number of moves needed to transfer n disks. In the simplest possible case $n = 1$, and obviously $H(1) = 1$.

If $n = 2$, transfer the top disk off the pile, move the bottom disk, and then transfer the first disk back onto the bottom disk; thus

$$H(2) = 1 + 1 + 1$$

or, reflecting the process more exactly,

$$H(2) = H(1) + 1 + H(1).$$

A similar analysis shows that $H(3) = H(2) + 1 + H(2)$. The analysis suggests that for n disks on a post we can (a) transfer the top $n - 1$ disks to a second post, which requires $H(n - 1)$ moves, (b) move the bottom disk to the third post, and (c) using $H(n - 1)$ moves, transfer the stack of $n - 1$ disks onto the third post. Thus

$$H(n) = H(n - 1) + 1 + H(n - 1) = 1 + 2H(n - 1).$$

The function H is easily written in Logo:

```
TO H :N
   IF :N = 1 [ OUTPUT 1 ]
   IF :N > 1 [ OUTPUT 1 + 2*H(:N − 1) ]
END
```

Modeling the problem with a spreadsheet

Enter 1 in cell B1 and the formula $1 + 2*B1$ in cell B2. Replicate this formula down column B to obtain figure 18.3.

Using the models

Students can use either of these models to obtain any term in the sequence

$$1, 3, 7, 15, 31, 63, \ldots,$$

whose values are the number of moves required to solve the puzzle begin-

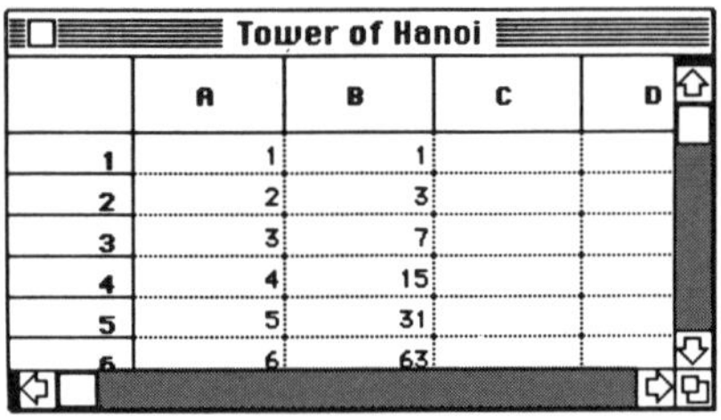

Fig. 18.3

ning with 1, 2, 3, 4, . . . disks. In particular, they can determine how many moves are needed to solve the commercial tower game, which has 7 disks. They can answer such questions as, "Mark has time to make 1000 moves. What's the biggest tower puzzle he could solve?" Thus they have a complete solution to the tower puzzle. Most students will notice that each number in the sequence is one less than a power of 2. Does this remain true for each term in the sequence? How can we be sure? A verification of this observation is presented later in the chapter.

A Zookeeper's Puzzle

The problem

A certain zookeeper has n cages in a row and two indistinguishable lions. The lions have to be in separate cages, and they may not be placed in adjacent cages. How many ways could the zookeeper assign lions to cages?

Modeling the problem with a Logo function

Let $L(n)$ denote the number of ways two lions can be placed into n cages. If $n = 2$, the task is impossible. Thus, in this simplest case, $L(2) = 0$.

If $n = 3$, the zookeeper could either put one lion in the first cage or not. If a lion is placed in the first cage, there is one acceptable way to place the second lion. If the first cage is not used, there are $L(2)$ ways to place the lions. Thus $L(3) = 1 + L(2)$.

Similarly, $L(4) = 2 + L(3)$, with the first term in the sum arising from those situations where one lion is placed in the first cage and the second term arising from arrangements leaving the first cage empty. This suggests that, in general $L(n) = (n - 2) + L(n - 1)$.

Written in Logo this function is

```
TO  L :N
    IF :N = 2 [ OUTPUT 0 ]
    IF :N > 2 [ OUTPUT (:N − 2) + L(:N − 1) ]
END
```

Modeling the problem with a spreadsheet

Enter 0 in cell B2 and the formula A3 − 2 + B2 in cell B3. Replicate this formula down column B to obtain figure 18.4

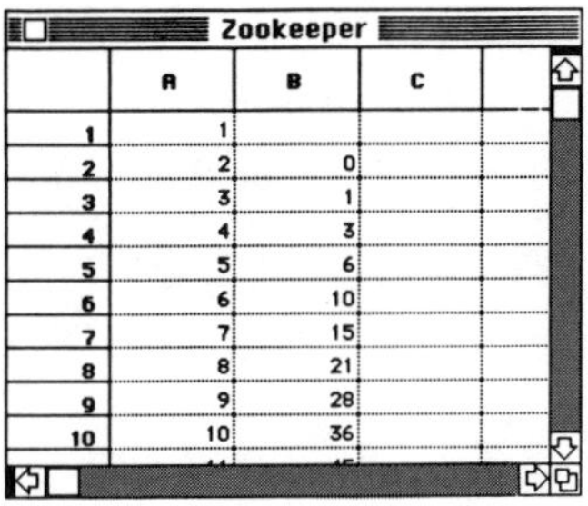

	A	B	C
1	1		
2	2	0	
3	3	1	
4	4	3	
5	5	6	
6	6	10	
7	7	15	
8	8	21	
9	9	28	
10	10	36	

Fig. 18.4

Using the models

Students can use one of the models to find $L(n)$ for any value of n. They can answer such questions as, "How many ways can two lions be placed in seven cages?" or, "If the zookeeper wants to place the lions differently each day for a month, how many cages are needed?" They may write down the sequence of values for L, that is, 0, 1, 3, 6, 10, 15, 21, 28, . . . and realize that this is the sequence of triangular numbers.

Annuity Holders' Bookkeeping

The problem

When Sonja deposits a series of equal annual payments in an account, she is creating an annuity. In a simplified example, if she deposits \$100 each year in an account that offers 9 percent yearly interest, what is the value of the annuity after a certain number of years?

Modeling the problem with Logo

The following chart shows how much money is in the annuity at the beginning of each year for three years:

Year	1	2	3
Amount	\$100	\$209 (\$100 + (1.09)\$100)	\$327.81 (\$100 + (1.09)\$209)

If $A(n)$ denotes the amount in the annuity at the beginning of the nth year, the recursive definition of A,

$$A(n) = \begin{cases} 100 & \text{if } n = 1 \\ 100 + (1.09)A(n - 1) & \text{if } n > 1, \end{cases}$$

can be expressed in Logo as follows:

```
TO A :N
   IF :N = 1 [ OUTPUT 100 ]
   IF :N > 1 [ OUTPUT 100 + 1.09*A(:N - 1) ]
END
```

Modeling the problem with a spreadsheet

Figure 18.5 shows the spreadsheet for this sequence.

Annuity

	A	B	C	D
1	1	$100.00		
2	2	$209.00		
3	3	$327.81		
4	4	$457.31		
5	5	$598.47		
6	6	$752.33		
7	7	$920.04		
8	8	$1102.85		
9	9	$1302.10		
10	10	$1519.29		
11	11	$1756.03		
12	12	$2014.07		

Fig. 18.5

Using the models

Typical questions students can use the models to answer are, "What is the value of the annuity at the beginning of year 10?" or, "How many years are needed for the value of the annuity to exceed $3000?" Even more important, students familiar with the use of models of situations may pose more interesting questions, such as, "How much should Sonja deposit annually in order to have $5000 in the annuity after fifteen years?" They can answer such a question by successive approximation of the necessary annual deposit. Consult chapter 19 for additional information about spreadsheets.

Questions about annuities can be posed and answered by students who know only a recursive rule; they do not need to know a closed-form rule. It is relatively easy for students to interpret a model of a situation with a Logo procedure or a spreadsheet because the syntax is so like the corresponding algebraic expression.

RECURSIVE THINKING IN APPLICATIONS

Students with experience in formulating mathematical ideas recursively have an advantage when they learn many of the applications of mathematics. This payoff is illustrated by a few examples below.

Exponential Growth

A country has a population of $P(t)$ people at time t, where t is the number of years since the country's founding. The population is 15 000 000 when $t = 0$, and the population increases at a rate of 3 percent a year. This population can be modeled with the following recursive function:

$$P(t) = \begin{cases} 15\ 000\ 000 & \text{if } t = 1 \\ P(t - 1) + 0.03P(t - 1) & \text{if } t > 1 \end{cases}$$

This is the well-known model of discrete exponential growth.

Logistic Growth

Suppose that the country has a ceiling population of 22 000 000 people and that growth is slowed as this maximum is approached. The previous model could be adapted to this situation by replacing the constant growth rate, 0.03, with the variable growth rate $0.03(1 - \dfrac{P(t - 1)}{22\ 000\ 000})$. The factor $(1 - \dfrac{P(t - 1)}{22\ 000\ 000})$ is close to 1 when $P(t - 1)$ is small, giving a growth rate that is near 0.03, and it is close to 0 when $P(t - 1)$ is close to 22 000 000, giving a near-zero growth rate. The following is a recursive function that can be used to model such a population:

$$P(t) = \begin{cases} 15\ 000\ 000 & \text{if } t = 1 \\ P(t - 1) + 0.03P(t - 1)\,(1 - \dfrac{P(t - 1)}{22\ 000\ 000}) & \text{if } t > 1 \end{cases}$$

Newton's Law of Cooling

Consider studying an object that is cooling, such as a just-fired piece of pottery. Newton assumed that the change in temperature of an object during a fixed time interval is proportional to the difference of the temperature between the object and its surrounding environment. The change in temperature is $T(t) - T(t - 1)$ and the temperature difference between the object and its surroundings is $S - T(t - 1)$, where S is the fixed outside temperature. Newton's assumption then translates into

$$T(t) - T(t - 1) = r \cdot (S - T(t - 1)),$$

or

$$T(t) = T(t - 1) + r \cdot (S - T(t - 1)),$$

where r is the constant of proportion.

A question students can understand and study is "If a kiln, originally at 1000°F, cools to 500°F in five hours in a surrounding temperature of 70°F, when will it reach a temperature of 250°F?" Students will first need to make a decision about the size of the time increment. Then they can study how that choice affects their answers.

NEW UNDERSTANDING OF PROOF BY INDUCTION

A typical example of a theorem to prove by mathematical induction is

The sum of the squares of the first n integers is
n(n + 1)(2n + 1)/6, for any positive integer n.

The proof depends on the principle of induction, a property of the set of positive integers N.

Let B be a subset of N with the following two conditions:

1. $1 \in B$.
2. If $n - 1 \in B$, then $n \in B$, for any $n > 1$.

Then $B = N$.

To apply the principle, one chooses for B the set of integers n for which the statement "The sum of the squares of the first n integers is $n(n + 1)$ $(2n + 1)/6$" is true. Verifying condition 1 is easy, and verifying condition 2 requires some straightforward algebraic manipulation. The hardest part for students, however, is the invocation of the induction principle. Often the proof is unconvincing because of its seeming circularity.

If the theorem above is restated using functional notation, however, a much simpler and more accessible proof becomes available. This proof does not explicitly call on the principle of induction. Here is the restatement:

$F(n) = G(n)$ for any positive integer n, where $F(n) =$ the sum of the squares of the first n integers and $G(n) = n(n + 1)(2n + 1)/6$.

The key to proceeding further is to rewrite the function F recursively:

$$F(n) = \begin{cases} 1 & \text{if } n = 1 \\ n^2 + F(n - 1) & \text{if } n > 1 \end{cases}$$

To show that F and G are the same function, we need only show that G satisfies the definition of F. The algebra required for this is the same as in the traditional proof, and with that the proof is complete. The set B was not constructed, nor was there mention of such a potentially misleading statement as, "Assume the theorem proved for n, then. . . ."

A second example of this method comes from the question raised at the end of the discussion of the Tower of Hanoi problem. There we considered the sequence

$$1, 3, 7, 15, 31, 63, \ldots,$$

of values of the function H defined recursively by

$$H(n) = \begin{cases} 1 & \text{if } n = 1 \\ 1 + 2H(n - 1) & \text{if } n > 1. \end{cases}$$

Is this sequence the same as that given by the function G defined in closed form by $G(n) = 2^n - 1$ for $n \geq 1$?

Computing both sequences in a spreadsheet will verify that $H(n) = G(n)$ for as many values of n as our spreadsheet will handle. To prove that $G(n) = H(n)$ for *all* positive n using induction, it's enough to show that $G(1) = 1$ and that $G(n) = 1 + 2G(n - 1)$ for $n \geq 2$. The first of these requirements is trivial; the second is merely a restatement of the identity $2^n - 1 = 1 + 2 (2^{n-1} - 1)$, and with that restatement the proof is complete.

Of course, the principle of induction *was* used implicitly, and it is worthwhile to see where it appeared. In fact, the very act of defining a function recursively requires the principle of induction. How do we know that the function F used in our first example actually has the set of all positive integers as domain? If we let B denote this domain, then surely $1 \in B$, and further, whenever F is defined for an argument $n - 1$, then it is defined for n, using the defining statement for F; thus the domain B is indeed N. The verification that H from our second example does indeed have N as its domain is entirely similar.

Burying the principle of induction in the definition of recursive functions strikes the right pedagogical balance between rigor and intuition. First, it is an advantage to remove the explicit invocation of the principle from every inductive proof. Second, those comfortable with recursively defined functions will have little difficulty relating the defining statements for such functions to other situations, which is all that is necessary.

We are not proposing teaching proof by induction to first-year algebra students. We do think that greater experience in two areas important from the discrete mathematics perspective should be given to beginners: practice with recursive descriptions, and experience with the concept of a function. For more on induction, see chapter 21.

CONCLUSION

For secondary school students and teachers, problems best described recursively need not be postponed or avoided. We have the tools to create discrete models and to use them to answer interesting questions. The concept of a recursively defined function underlies these models, and this concept is accessible even to young students. The *Standards* notes that some traditional skills are being made obsolete by technology. Here we see some long-known mathematics that technology has made accessible and newly significant.

REFERENCE

National Council of Teachers of Mathematics. *Curriculum and Evaluation Standards for School Mathematics*. Reston, Va.: The Council, 1989.

19

Using Spreadsheets to Introduce Recursion and Difference Equations in High School Mathematics

Bruce R. Maxim
Roger F. Verhey

DIFFERENCE equations are valuable tools for working with discrete mathematical models (Sandefur 1985). When a spreadsheet template is set up to solve the difference equations, students can focus their attention on the analysis of the model and are not distracted by the task of solving the difference equations. This situation allows students to work on more realistic applications and to attack more interesting problems.

The study of difference equations leads quite naturally to the consideration of recursion and recursive functions. For example, $F(n) = n * F(n - 1)$ could be thought of as a difference equation because the value of $F(n)$ is n times the value of $F(n - 1)$. However, another way to think about this equation is to consider it a recursive definition for some function F, that is, a definition in which the function F is defined in terms of itself. To complete the definition of F, or any other recursive function, the value of $F(n)$ must be known for a particular value of n. For example, it might be that $F(1) = 1$. Combining this statement with the earlier one gives a complete definition for the function F as follows:

$$F(n) = \begin{cases} 1 & \text{if } n = 1 \\ n * F(n - 1) & \text{if } n > 1 \end{cases}$$

This is a recursive definition for the factorial function, namely, $F(n) = n!$.

There are several reasons for wanting to examine recursive functions using a spreadsheet program. First, each step of evaluating a recursive function can be observed visually. Second, spreadsheet templates set up to allow the evaluation of recursive functions are very efficient in their use of computer memory. Third, creating a spreadsheet template to evaluate a recursive function can offer unique opportunities for class discussion regarding the nature and efficiency of recursion. Finally, evaluating a recursive function using a spreadsheet template can give students an appreciation of the re-calculation process used by the spreadsheet program.

The formulas for a spreadsheet template to evaluate the factorial function are shown in figure 19.1. The entries in column A are formulas for computing values for n, and the entries in column B are formulas for computing the corresponding values of $F(n)$. The first step in building the template is to place the constant 1 in cells A2 and B2. This handles the situation $F(1) = 1! = 1$. The cells A3 and B3 then handle the situation $F(2) = 2! = 2 * F(1)$. Thus, the entry in cell A3 is A2 + 1, and the entry in cell B3 is A3 * B2. To compute larger factorials, replicate the expressions from cells A3 and B3 into the respective cells A4, B4, A5, B5, and so on, using the relative reference option of the spreadsheet program. Continue replicating these formulas until the desired value of n appears in column A. When the formulas are evaluated, the corresponding entry in column B will contain the value of $n!$.

	A	B
	n	F(n) = n!
1		
2	1	1
3	A2 + 1	A3*B2
4	A3 + 1	A4*B3
5	A4 + 1	A5*B4

Fig. 19.1

The spreadsheet template shown in figure 19.2 displays the values of n and $n!$ for $n = 1$ to 4. A spreadsheet template to compute 5! would contain 5 rows, 6! would contain 6 rows, and so on. The model will need to be extended each time $n!$ is to be evaluated for a larger value of n.

A spreadsheet template for the factorial function that can calculate the value of the factorial function for any value of n is given in figure 19.3. This template uses the manual recalculation option of the spreadsheet program.

	A	B			A	B
	n	F(n) = n!			N	N!
1				1		
2	1	1		2	A2 + 1	IF(A2 = 1,1,A2*B2)
3	2	2				
4	3	6				
5	4	24				

Fig. 19.2   Fig. 19.3

Initially, the constant 0 is entered into cell A2 to give it a value before entering the formula A2 + 1 in the same cell. The value displayed in cell A2, after entering the formula A2 + 1, will then be 1.

Since the definition of $F(n)$ depends on the value of n, the spreadsheet program's logical IF function needs to be used. This allows for the initial condition $F(1) = 1$ as well as the more general case $F(n) = n * F(n - 1)$. The formula entered in cell B2 is IF(A2 = 1, 1, A2 * B2). This means that if the value of cell A2 is 1, then the value of cell B2 will be 1; otherwise, the value of cell B2 will be the product of the current values of cells B2 and A2. Note that the recalculated value of cell B2 depends on the current value of B2. Thus, the initial value displayed in cell B2 will be 1, which is the value of 1!. To compute successive values of $n!$, use the manual recalculation feature of the spreadsheet program repeatedly until cell A2 contains the desired value of n. It should be noted that the entry of 0 in cell A2, followed by the entry of the formula A2 + 1, must be done before entering the IF function call into cell B2.

These two examples might be given to students as part of an introduction to the concept of recursion. The class should carefully examine the structure of each template, paying close attention to the formulas in each cell. Class discussion should consider the amount of time it takes to define each template, the effort required to use each, and the amount of effort needed to revise each template to compute new values of $n!$. The students might then be encouraged to devise spreadsheet templates to evaluate other functions that have recursive definitions (Arganbright 1984; Maxim and Verhey 1988).

An example of a spreadsheet template to study a first-order linear difference equation is shown in figure 19.4. This template computes population growth for a period of seven years. It shows an initial population, $A(0)$, of 1000 creatures and an annual growth rate, r, of 6 percent.

One model for population growth assumes that growth during a given time period is proportional to the size of the population at the beginning of the time period. The difference equation to show this relationship is

$$A(n + 1) - A(n) = r * A(n),$$

which is a first-order linear difference equation. To facilitate the development of the spreadsheet template, the equation is rewritten as

$$A(n + 1) = (1 + r) * A(n).$$

The behavior of this equation should be developed using class discussion. The discussion might be initiated by considering the size of the population at the end of year 1, $[A(1) = (1 + r) * A(0)]$, then at the end of year 2, $[A(2) = (1 + r) * A(1)]$, and then in general—discussing how each new value is obtained from a previously known value. As the discussion turns toward the development of the spreadsheet template, the students need to be confident that they understand how the initial population $A(0)$ and the growth rate r relate to the computation of $A(n)$. In the spreadsheet shown in figure 19.4, the value of $A(0)$ is stored as a constant in cell B11 and the value of r is stored as a constant in cell C12. Cell B12 contains the expression (1 + C12) * B11, cell B13 contains the expression (1 + C12) * B12, and so on. The size of the population at the end of the seventh year appears in cell B18.

	A	B	C	D
1				
2		First-Order Difference Equation		
3		Population Growth		
4				
5		A(n) is size of population at time n.		
6				
7		A(n + 1) − A(n) = r * A(n)		
8		or		
9		A(n + 1) = (1 + r) * A(n)		
10				
11	A(0) =	1,000.00	r	
12	A(1) =	1,060.00	.06	
13	A(2) =	1,123.60		
14	A(3) =	1,191.02		
15	A(4) =	1,262.48		
16	A(5) =	1,338.23		
17	A(6) =	1,418.52		
18	A(7) =	1,503.63		

Fig. 19.4

The students should be encouraged to experiment with various values for $A(0)$ and r, and be challenged to answer questions like "What happens to the size of the population after seven years if the initial population size is doubled?" and "How long does it take for the population to double in size when the growth rate is 6 percent?" The beauty of using a spreadsheet is that the students need only enter a new value for $A(0)$ in cell B11 and a new value for r in cell C12, and they can watch the spreadsheet recalculate all other numerical values. The spreadsheet's strength as a problem-solving tool is further illustrated by considering the question, "What rate, r, will cause the population to double in seven years?" In this situation, the investigator enters different values for r into cell C12 until the value in cell B18 is twice the value in cell B11. Students receive immediate feedback on the results of their experiments in the form of visual records. Moreover, the results of each experiment can easily be copied to the system printer.

Ensuing class discussion might lead to whether or not biological growth can continue in such an unbounded manner. The answer is of course that it cannot. Ultimately, the population will grow too large for the available food supply. The class might be challenged to devise a difference equation that takes this into account (Sandefur 1985; Maxim and Verhey 1987).

A spreadsheet program is a very good problem-solving tool when the desired result is known and the investigator wants to determine an initial condition. Our next two examples illustrate this problem-solving capability of spreadsheet programs. First, consider the amortization of a loan. The spreadsheet template shown in figure 19.5 exhibits a stage in the analysis of the question, "How much can a person borrow from the bank at a 12 percent interest rate and repay with twenty-four monthly payments of $250?" This is an important question, and it is frequently asked by borrowers to determine the maximum amount of money that they can borrow, given their ability to repay the loan. To use the spreadsheet in figure 19.5 to answer this question, enter different values in cell D6 for the size of the initial loan, $A(0)$, until the value computed for A(24) in cell D24 is zero.

	A	B	C	D
1		First-Order Difference Equations		
2		Loan Amortization		
3				
4	Monthly interest rate		MR	.01
5	Monthly payment		MP	250.00
6	Loan amount		A(0)	5,500.00
7	Value of loan at nth month		A(n)	
8				
9		$A(n + 1) = (1 + MR) * A(n) - MP$		
10				
11	n	A(n)	n	A(n)
12				
13	1	5,305.00	13	2,807.16
14	2	5,108.05	14	2,585.23
15	3	4,909.13	15	2,361.08
16	4	4,708.22	16	2,134.69
17	5	4,505.30	17	1,906.04
18	6	4,300.35	18	1,675.10
19	7	4,093.35	19	1,441.85
20	8	3,884.28	20	1,206.27
21	9	3,673.12	21	968.33
22	10	3,459.85	22	728.01
23	11	3,244.45	23	485.29
24	12	3,026.89	24	240.14

Fig. 19.5

The construction of the template in figure 19.5 is similar to the construction of the one in figure 19.4. However, notice that the interest rate is monthly because the payments are made on a monthly basis. The value displayed in cell B13 is the value of $A(1)$, which is the value of the loan after the first monthly payment has been made. It may be helpful to discuss one of the formulas in constructing the template. The formula in cell D13, $(1 + D4) * B24 - D5$, gives the value of $A(13)$, where B24 is the value computed for $A(12)$. For this template the recalculation feature must be set to calculate by columns rather than rows. A similar method can be used to find the monthly payment for a loan of a specific amount.

To solve a second-order linear difference equation, the class might develop as an example a spreadsheet template to solve the Gambler's Ruin problem (Feller 1950). This problem is not usually seen in secondary school mathematics because the solution using traditional methods is difficult. Yet it is an interesting problem and quite approachable using the technique described below (Sandefur 1985).

In this problem a person bets $1 on each play of a game. The probability of winning on each play might be 0.4, and the probability of losing would then be 0.6. A person plays the game until either going broke or winning a set amount of money, perhaps $10. What is the probability of going broke if a person begins playing with $5?

To express this problem mathematically we let $P(n)$ be the probability of eventually going broke if a person presently has n dollars. We then consider the situation when a person has $n + 1$ dollars. There are two ways in which this person can go broke. Either the player can win a dollar and then go broke after possessing $n + 2$ dollars or the player can lose a dollar and go broke after possessing n dollars. This situation can be described with a tree diagram as shown below.

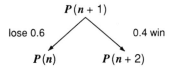

Since the probability of winning a dollar on a single play of the game is 0.4, the probability that a player starting with $n + 1$ dollars will go broke after possessing $n + 2$ dollars is obtained by following the right branch of the tree. Using the multiplication rule for conditional probabilities, we find that the probability of winning a dollar and then going broke is $0.4 * P(n + 2)$. By contrast, the probability of losing a dollar and then going broke is $0.6 * P(n)$, which is obtained by following the left branch of the tree. These two events are mutually exclusive, and we may use the addition rule to combine the probabilities. Thus a mathematical expression for the value of $P(n + 1)$ is

$$P(n + 1) = 0.4 * P(n + 2) + 0.6 * P(n).$$

This equation may be rewritten as

$$P(n + 2) = 2.5 * P(n + 1) - 1.5 * P(n),$$

which is a second-order linear difference equation, since the difference between time $n + 2$ and time n is 2.

The fact that we are working with a second-order difference equation makes it difficult to construct a template to model this problem. To compute the value of $P(2)$, we need to have values for both $P(1)$ and $P(0)$. The value of $P(0)$ is known to us, since it is one of the boundary conditions, and its value is 1. If we reach the goal, we quit playing and cannot go broke, thus $P(10) = 0$. But there is nothing in the problem statement that gives us a value for $P(1)$. Moreover, the difference equation does not allow us to compute $P(10)$ without first having a value for $P(1)$. Thus the spreadsheet template we construct for this problem will need to allow us to search for a value of $P(1)$ that causes the value of $P(10)$ to be computed as 0.

One way to handle this problem is to construct the template assigning an arbitrary value for $P(1)$ less than 1, since $P(1)$ is a probability. The template shown in figure 19.6 shows the boundary value of 1 entered for $P(0)$ in cell E5 and an initial value of 0.99 entered for $P(1)$ in cell E6. The difference equation can then be entered into cells E7 through E15, which calculate values for $P(2)$ through $P(10)$, respectively. For example, the expression $2.5 * E6 - 1.5 * E5$ is entered into cell E7. This expression computes the value of $P(2)$. This expression is then replicated in cells E8 through E15 using the spreadsheet program's replicate function and its relative referencing feature.

	A	B	C	D	E
1			The Gambler's Ruin		
2			P(n) is the probability of going broke starting with $n		
3			P(n + 2) = 2.5 * P(n + 1) − 1.5 * P(n)		
4					
5	the value of P(0) is				1.000000000
6	If the value of P(1) is				.990000000
7	the value of P(2) is				.975000000
8	the value of P(3) is				.952500000
9	the value of P(4) is				.918750000
10	the value of P(5) is				.868125000
11	the value of P(6) is				.792187500
12	the value of P(7) is				.678281250
13	the value of P(8) is				.507421875
14	the value of P(9) is				.251132813
15	the value of P(10) is				− .133300780

Fig. 19.6

Once the template has been constructed, the students can experiment by entering different values for $P(1)$ in cell E6. The goal is to find a value for $P(1)$ that yields a value for $P(10)$ that is close to 0. An integral part of this exploration is a discussion of what it means to be "close enough" to 0. The template shown in figure 19.7 computes a value for $P(10)$ that is within three-thousandths of 0, when the value used for $P(1)$ is 0.9912. The value of $P(5)$ shown in figure 19.7 is 0.88395. This represents the probability of a player going broke after starting with $5. This value is very close to the value obtained by Sandefur (1985). It is interesting to note that the spreadsheet shown in figure 19.7 also computes values for $P(n)$ for $n = 1$ to 9. The students might be surprised to find that the probability of going broke even when starting with $8 is greater than 0.5.

The class may want to explore the problem further using different probabilities for winning and losing on each play of the game. The template shown in figure 19.7 could easily be modified to allow these probabilities to be entered as parameters in the template. The students might also be interested in comparing the results of the spreadsheet model to those that can be obtained from a computer program that uses the Monte Carlo method to solve the problem by simulation (Verhey and Maxim 1987).

Spreadsheet programs give students a valuable tool to explore key concepts and applications of discrete mathematics at the high school level. The programming demands made on the students are not as great as those made by programming languages such as BASIC or Pascal. Spreadsheets give greater flexibility and can be less tedious to use than calculators. Working with spreadsheet programs can promote cooperative problem-solving skills in students encouraged to work in pairs as they develop templates to solve interesting problems.

	A	B	C	D	E
1			The Gambler's Ruin		
2		P(n) is the probability of going broke starting with $n			
3		P(n + 2) = 2.5 * P(n + 1) − 1.5 * P(n)			
4					
5	the value of P(0) is				1.000000000
6	If the value of P(1) is				.991200000
7	the value of P(2) is				.978000000
8	the value of P(3) is				.958200000
9	the value of P(4) is				.928500000
10	the value of P(5) is				.883950000
11	the value of P(6) is				.817125000
12	the value of P(7) is				.716887500
13	the value of P(8) is				.566531250
14	the value of P(9) is				.340996875
15	the value of P(10) is				.002695313

Fig. 19.7

REFERENCES

Arganbright, Deane. *Mathematical Applications of Electronic Spreadsheets*. New York: McGraw-Hill Book Co., 1984.

Feller, William. *An Introduction to Probability Theory and Its Applications*, vol. 1. New York: John Wiley & Sons, 1950.

Maxim, Bruce R., and Roger F. Verhey. "Spreadsheets in Mathematics: Linear and Non-Linear Difference Equations." *MACUL Newsletter*. [Michigan Association of Computer Users in Learning] 6, no. 16 (April/May 1987): 7.

———. "Spreadsheets in Mathematics: Recursive Algorithms." *MACUL Newsletter* [Michigan Association of Computer Users in Learning] 8, no. 4 (March 1988): 17.

Sandefur, James T. "Discrete Mathematics: A Unified Approach." In *The Secondary School Mathematics Curriculum*, 1985 Yearbook of the National Council of Teachers of Mathematics, edited by Christian R. Hirsch. Reston, Va.: The Council, 1985.

Verhey, Roger F., and Bruce R. Maxim. "Spreadsheets in Mathematics: The Gambler's Ruin Problem." *MACUL Newsletter* [Michigan Association of Computer Users in Learning] 6, no. 17 (June 1987): 8–9.

20

Recursive Thinking: A Method for Problem Solving

Sandra Burrell **Estelle LeGrand**
Mildred Eaton **Darlene Morris**
Estelle Feeling **Gertie Tillerson**
Fauzy Ghareeb **Robert Williams**

A S A method for problem solving, recursive thinking can be used effectively at several levels of mathematics. In the spirit of the *Curriculum and Evaluation Standards for School Mathematics* (NCTM 1989), we are presenting one problem with several approaches to demonstrate how solving problems recursively can span mathematics from junior high school to college. The problem presented below was chosen because it represents a real-life situation with which most students can readily identify.

Suppose that (insert a popular movie or rock band) is playing at the Star Theater. This theater has one section. The seats in the section are arranged so that there are 70 seats in the first row, 72 seats in the second row, 74 seats in the third row, 76 seats in the fourth row, and so on, for a total of 30 rows. The seats are numbered from left to right. The first seat in the first row is number 1, the first seat in the second row is number 71, the first seat in the third row is number 143, ... Your friend bought you a ticket for seat 1000. Seats near the center of the theater are considered good seats. Some possible questions are the following:

1. How many seats are in the last row?
2. How many seats are in the theater?
3. Do you have a good seat?

Warm-up. Let n represent the row number, $n = 1, \ldots, 30$. Let $R(n)$ be the number of seats in row n. Thus, $R(1) = 70$, $R(2) = 72$, $R(3) = 74$,

This paper was written as a final project for the 1989 NSF-funded summer workshop, Modeling with Discrete Mathematics, which was held at Georgetown University. We would like to thank the project directors, James Sandefur and Monica Neagoy, for their helpful suggestions in the preparation of this paper.

and so forth. In general, row n has two more seats than row $n - 1$. This can be expressed by the relation

$$R(n) = R(n - 1) + 2, \text{ for } n > 1. \tag{1}$$

Have the students use this recursive relationship to find the number of seats in each of the thirty rows, possibly using a calculator.

Presentation: *For all levels.* From table 20.1, we see that a direct computation for the number of seats in each row is

$$R(n) = 70 + 2(n - 1) = 68 + 2n. \tag{2}$$

Thus the number of seats in each row can be determined recursively using formula 1 or directly using formula 2. Notice that the direct computation made it easier to compute $R(30)$, but the recursive formulation is the way most people begin such problems; that is, when people see the sequence of numbers

$$70, 72, 74, \ldots ,$$

they usually think first "each number is 2 more than the previous number," not "the nth number is $68 + 2n$." The direct computation comes from further study of the sequence.

TABLE 20.1

n	$R(n)$	
1	$70 = 70$	$= 70 + 0(2)$
2	$72 = 70 + 2$	$= 70 + 1(2)$
3	$74 = 70 + 2 + 2$	$= 70 + 2(2)$
4	$76 = 70 + 2 + 2 + 2$	$= 70 + 3(2)$
	$\vdots$	
n	$R(n) = 70 + 2 + \ldots + 2 = 70 + (n - 1)(2)$	
	$\vdots$	
30	$128 = 70 + 2 + \ldots + 2$	$= 70 + 29(2)$

Now let us approach the problem of determining the *total* number of seats in the first n rows. Denote this number by $T(n)$. Thus $T(1) = 70$, $T(2) = 70 + 72 = 142$, $T(3) = 142 + 74 = 216$, and so on.

First approach: *For all levels.* The total number of seats in the first n rows is the total number of seats in the first $n - 1$ rows plus the number of seats in row n, as shown in table 20.2. You might want to omit the last row in the table and instead use the table to generate $T(n)$ for $n = 5, 6, \ldots$, possibly using a calculator. Each value of $T(n)$ is obtained by adding the number above it in the column to the number to the left of it in the first column.

We are using a recursive method to find $T(n)$; that is,

$$T(n) = T(n - 1) + R(n)$$

TABLE 20.2

n	$R(n)$	$T(n)$
1	$R(1) = 70$	$T(1) = R(1) = 70$
2	$R(2) = 72$	$T(2) = T(1) + R(2) = 142$
3	$R(3) = 74$	$T(3) = T(2) + R(3) = 216$
4	$R(4) = 76$	$T(4) = T(3) + R(4) = 292$
.		
.		
.		
n	$R(n) = 68 + 2n$	$T(n) = T(n-1) + R(n) = T(n-1) + 68 + 2n$

with $T(1) = 70$, but this method is time-consuming. We would like to develop a direct computation for $T(n)$.

Second approach: *For geometry students.* The horizontal line in figure 20.1 represents 70 seats, and additional seats are sketched beside the line.

$$
\begin{aligned}
T(1) &= R(1) & &= 70 & &= 70 + 2(0) \\
T(2) &= R(1) + R(2) & &= 70 + (70 + 2) & &= 2(70) + 2(1) \\
T(3) &= R(1) + R(2) + R(3) & &= 70 + (70 + 2) + (70 + 2 + 2) & &= 3(70) + 2(3) \\
T(4) &= R(1) + \ldots + R(4) & &= 70 + \ldots + (70 + 2 + 2 + 2) & &= 4(70) + 2(6) \\
T(5) &= R(1) + \ldots + R(5) & &= 70 + \ldots + (70 + 2 + 2 + 2 + 2) & &= 5(70) + 2(10)
\end{aligned}
$$

Fig. 20.1. Diagram of seats in the theater

The numbers 0, 1, 3, 6, and 10 shown on the diagram and as factors of the last term of the equations in figure 20.1 are the triangular numbers. Recall (or establish) that the nth triangular number equals

$$
\frac{n(n-1)}{2} \ .
$$

From figure 20.1, we see that the same number of seats is added at both ends of each row. Thus, the total number of additional seats for n rows is

$$
2\left(\frac{n(n-1)}{2}\right) \ .
$$

Thus, we see that

$$
T(n) = n(70) + 2\left(\frac{n(n-1)}{2}\right) = n^2 + 69n.
$$

Substituting $n = 30$, we find that the total number of seats in the theater is

$$
T(30) = 30^2 + 69(30) = 2970.
$$

Third approach: *For algebra students.* To find the total number of seats in the first n rows, sum $R(1) + \ldots + R(n)$.

$$R(1) = R(1) + 2(0)$$
$$R(2) = R(1) + 2(1)$$
$$R(3) = R(1) + 2(2)$$
$$\vdots$$

$$\underline{R(n) = R(1) + 2(n-1)}$$
$$T(n) = nR(1) + 2(0 + 1 + \ldots + (n-1))$$

and again use (or prove) the fact that

$$1 + 2 + \ldots + (n-1) = \frac{n(n-1)}{2}$$

and that $R(1) = 70$ to get

$$T(n) = n(70) + 2\left(\frac{n(n-1)}{2}\right) = n^2 + 69n.$$

Fourth approach: *For precalculus or calculus students.* Here we deal directly with the recursive relation

$$T(n) = T(n-1) + R(n) = T(n-1) + 68 + 2n.$$

When we were counting the number of seats in each row, we had the recursive relation and the corresponding direct computation formula,

$$R(n) = R(n-1) + 2 \quad \text{and} \quad R(n) = 68 + 2n,$$

respectively. Using the direct computation, we find that $R(n-1) = 68 + 2(n-1) = 66 + 2n$. Substitution of the direct computation for $R(n)$ and $R(n-1)$ into the recursive formula gives

$$68 + 2n = (66 + 2n) + 2,$$

which is balanced.

When each number is the previous number plus a constant, then the direct computation involved an n term. For the total number of seats, each number is the previous number plus a number involving n. We might then expect that the direct computation involves an n^2 term; that is,

$$T(n) = an^2 + bn + c,$$

for some constants, a, b, and c. In this case, $T(n-1) = a(n-1)^2 + b(n-1) + c$. Substitution of the direct computation formula into the recursive relation should give equality; that is,

$$(an^2 + bn + c) = (a(n-1)^2 + b(n-1) + c) + 68 + 2n.$$

Simplifying this equation and collecting all terms on the right will give

$$0 = (2 - 2a)n + a - b + 68.$$

We have equality if $2 - 2a = 0$ and $a - b + 68 = 0$. But this means that $a = 1$ and $b = 69$. Thus, the closed-form formula must be

$$T(n) = n^2 + 69n + c.$$

Since $T(1) = 70$ and our formula gives $T(1) = 1^2 + 69(1) + c$, setting the right sides equal to each other and solving gives $c = 0$. The total is

$$T(n) = n^2 + 69n$$

as before.

Now, let us address the question, In which row is seat number 1000?

First approach: *For all levels.* We compute $T(1) = 70, \ldots, T(12) = 972$, $T(13) = 1066$. (You could do these computations using either the recursive or the closed-form formula.) Thus, seat number 1000 is in row 13. The seats in row 13 are numbered 973 through 1066. Thus our seat is the 28th seat out of 94. The middle seats are the 47th and 48th, so we are nineteen seats from the middle.

Second approach: *For upper-level students.* Using the closed-form formula, we want a value of n that makes $T(n) > 1000$. Letting $T(n) = 1000$ and solving $1000 = n^2 + 69n$ with the quadratic formula, we find that

$$n = \frac{-69 \pm \sqrt{(69)^2 + 4000}}{2} = 12.3, \text{ or } -81.3.$$

Thus, if $n \geq 13$, then we have more than 1000 seats. The 1000th seat is in row 13. We can use the previous approach to find the exact position.

Many other interesting problems that can be solved using some of the techniques presented are in Seymour and Shedd (1973). For a more indepth study of recursive relations at a higher level, see Sandefur (1990).

REFERENCES

National Council of Teachers of Mathematics. *Curriculum and Evaluation Standards for School Mathematics.* Reston, Va.: The Council, 1989.

Sandefur, James T. *Discrete Dynamical Systems: Theory and Applications.* Oxford, England: Oxford University Press, 1990.

Seymour, Dale, and Margaret Shedd. *Finite Differences.* Palo Alto, Calif.: Dale Seymour Publications, 1973.

21

Teaching Induction in Discrete Mathematics

Kay E. Smith

T HE NCTM *Curriculum and Evaluation Standards for School Mathematics* recommends giving increased attention to proof by mathematical induction, including proofs in contexts other than series (NCTM 1989, pp. 143, 145). The *Standards* also advocates experience in thinking recursively and in formulating conjectures by generalizing from particular cases (pp. 143, 177). The purpose of this chapter is to illustrate the use of these methods in discrete mathematics and, in particular, to demonstrate how formulating conjectures and thinking recursively can aid the teaching of mathematical induction.

For most students, summation formulas are their first experience with mathematical induction. The proofs of these formulas are an important application of mathematical induction, but students can learn to do them and still have little understanding of how to apply mathematical induction in other contexts. In their article, "Guessing, Mathematical Induction, and a Remarkable Fibonacci Result," Crawford and Long (1978, p. 613) observe that students who are learning mathematical induction

> tend to confuse the logic of the proof with the manipulation required to deal with the finite sums. Somehow, they come to think, a proof by mathematical induction always involves adding the same term to both sides of an equation that was assumed to be true in the first place!

The author's experience corroborates this observation. After learning how to apply mathematical induction to verify summation formulas, many students try to find an equation to manipulate whenever they are instructed to prove a statement by mathematical induction.

Crawford and Long suggest that to improve students' understanding and motivation, one should "introduce mathematical induction in a context rich in intrinsic interest" (p. 613). Combinatorics and graph theory provide many interesting applications of mathematical induction. Some of these are presented in this chapter.

Crawford and Long also suggest spending "a certain amount of time on activities designed to lead the students to guess the results they subsequently prove" (p. 613). In the examples below we illustrate a method for teaching mathematical induction that incorporates this suggestion. We pose a problem that depends on an integer variable n for students to investigate. To become familiar with the problem, students work out the solution for a few small values of n. They then look for patterns and make a conjecture about the answer to the problem. As they look for patterns, we encourage them to think recursively, that is, to determine how the solution for $n = k + 1$ is related to, or can be constructed from, the solution for $n = k$. When the students discover this relationship, they are prepared to write a proof of their conjecture by mathematical induction.

Principle of Mathematical Induction

In the examples in this chapter we use the following form of the principle of mathematical induction:

Let $P(n)$ be a statement that depends on the positive integer n, and let m be a positive integer. If
  (i) *(Basis step)* $P(m)$ is true and
  (ii) *(Inductive step)* for all positive integers $k \geq m$, $P(k + 1)$ is true if $P(k)$ is true,
then $P(n)$ is true for all positive integers $n \geq m$.

For a proof based on this principle, one must show that the two hypotheses are satisfied. To do so, first establish the basis step by verifying that the statement holds for the least positive integer m in which you are interested. For the inductive step assume that k is a positive integer greater than or equal to m and that the statement holds for k. Using this assumption, show that the statement is true for $k + 1$. After verifying the two hypotheses, you can invoke the principle of mathematical induction to conclude that the statement holds for all positive integers $n \geq m$.

Mathematical Induction in the Classroom

Example 1. How many subsets can be formed from $\{1, 2, \ldots, n\}$?

Solution. For small values of n, one can list all the subsets as shown in table 21.1. It appears that the number of subsets of $\{1, 2, \ldots, n\}$ is 2^n.

To prove this conjecture by mathematical induction, we need to find a relationship among the subsets for successive values of n. Observe that the subsets for $n = 2$ can be obtained by first listing the subsets of $\{1\}$ and then listing the sets formed by adjoining 2 to each of the subsets of $\{1\}$. Similarly, the subsets for $n = 3$ are the subsets of $\{1, 2\}$ and the subsets formed by adjoining 3 to each of the subsets of $\{1, 2\}$. These observations suggest that, in general, the subsets of $\{1, 2, \ldots, n + 1\}$ are the subsets of $\{1, 2, \ldots,$

TABLE 21.1

n	Subsets of $\{1, 2,..., n\}$	Number of subsets
1	$\phi, \{1\}$	2
2	$\phi, \{1\}, \{2\}, \{1, 2\}$	4
3	$\phi, \{1\}, \{2\}, \{1, 2\},$	8
	$\{3\}, \{1, 3\}, \{2, 3\}, \{1, 2, 3\}$	

$n\}$ and the subsets formed by adjoining $n + 1$ to each subset of $\{1, 2, \ldots, n\}$.

Having recognized the relationship among the subsets for successive integers, we are now ready to do a formal proof by mathematical induction.

THEOREM. *If n is a positive integer, the number of subsets of* $\{1, 2, \ldots, n\}$ *is* 2^n.

Proof. The set $\{1\}$ has two subsets, namely ϕ and $\{1\}$. Therefore the theorem is true for $n = 1$.

Assume that k is a positive integer and $\{1, 2, \ldots, k\}$ has 2^k subsets. Now consider the subsets of $\{1, 2, \ldots, k + 1\}$. A subset of $\{1, 2, \ldots, k + 1\}$ either contains $k + 1$ or it does not. If it does contain $k + 1$, then it can be formed by adjoining $k + 1$ to a subset of $\{1, 2, \ldots, k\}$. If it does not contain $k + 1$, then it is a subset of $\{1, 2, \ldots, k\}$. Conversely, all subsets of $\{1, 2, \ldots, k\}$ and all subsets formed by adjoining $k + 1$ to a subset of $\{1, 2, \ldots, k\}$ are subsets of $\{1, 2, \ldots, k + 1\}$. Therefore the number of subsets of $\{1, 2, \ldots, k + 1\}$ is twice the number of subsets of $\{1, 2, \ldots, k\}$, which by inductive assumption is 2^k. Hence the number of subsets of $\{1, 2, \ldots, k + 1\}$ is $2 \times 2^k = 2^{k+1}$. Thus the theorem is true for $k + 1$ if it is true for k. Therefore the theorem is true for all positive integers by the principle of mathematical induction.

In example 1, it was easy to guess the formula for the number of subsets by examining a few cases. We used recursive thinking to construct a proof of our conjecture. In the next example, we apply recursive thinking to discover both the desired formula and the key element in its proof.

Example 2. How many ways can $2n$ people be divided into n pairs?

Solution. For ease of notation, denote the people by $1, 2, \ldots, 2n$. The possible pairings for small values of n are shown in table 21.2. A formula relating the values in the first and last columns of the table may not be immediately obvious.

To find a formula for the number of pairings, consider how the pairings in table 21.2 were made. We obtained the pairings for $n = 2$, that is, four persons, by first selecting a partner for person number 1 and then pairing the remaining two people. Since there are three ways to select the partner

TABLE 21.2

n	$2n$	Pairings									Number of pairings
1	2	1-2									1
2	4	1-2	3-4,		1-3	2-4,		1-4	2-3		3
3	6	1-2	3-4	5-6,	1-2	3-5	4-6,	1-2	3-6	4-5,	15
		1-3	2-4	5-6,	1-3	2-5	4-6,	1-3	2-6	4-5,	
		1-4	2-3	5-6,	1-4	2-5	3-6,	1-4	2-6	3-5,	
		1-5	2-3	4-6,	1-5	2-4	3-6,	1-5	2-6	3-4,	
		1-6	2-3	4-5,	1-6	2-4	3-5,	1-6	2-5	3-4	

for person number 1 and one way to pair the remaining two people, the total number of pairings is 3×1. Similarly, to obtain the pairings for $n = 3$, we first selected a partner for person number 1 and then listed all ways to pair the remaining four individuals. Since there are five ways to select the partner for person number 1 and 3×1 ways to pair the remaining four individuals, the total number of pairings for six people is $5 \times 3 \times 1$.

Now the pattern is clearer. The number of ways to divide $2n$ people into n pairs appears to be the product of the first n odd integers—

$$(2n - 1) \times (2n - 3) \times \ldots \times 3 \times 1.$$

By generalizing the argument in the preceding paragraph, we could prove this conjecture by mathematical induction also.

Using a combinatorial argument, one can show that the number of ways to divide $2n$ people into n pairs is $(2n)!/(2^n n!)$. Students can be asked to prove by mathematical induction that the two formulas are equal.

The remaining examples are problems from graph theory. (For more details about graphs, see chap. 11.) Many graph-theory proofs using mathematical induction have the same basic structure. The variable is the number of vertices in the graph. At the inductive step, consider a graph with $k + 1$ vertices. By temporarily ignoring one of the vertices and the edges incident with it, obtain a subgraph with k vertices to which the inductive assumption can be applied. To complete the proof, show how the desired conclusion follows when the $(k + 1)$st vertex and the edges incident with it are added to this subgraph. The next example illustrates this approach.

Example 3. A *round robin tournament* is a tournament in which each competitor plays every other competitor exactly once. Assume that no ties occur. At the end of a round robin tournament with n players, is it always possible to rank the competitors so that the player ranked ith defeated the player ranked $(i + 1)$st for $i = 1, 2, \ldots, n - 1$?

Solution. A round robin tournament can be modeled by a directed graph in which the vertices represent the competitors and an edge is drawn from vertex i to vertex j if player i defeated player j. A path in this graph that includes every vertex exactly once corresponds to a ranking of the competitors in which the player ranked ith defeated the player ranked $(i + 1)$st.

A directed graph is *complete* if for every two vertices u and v, either (u, v) or (v, u) is an edge of the graph. If we assume that no ties occur, the graph model of a round robin tournament is complete. Hence in graph theory terms the question we seek to answer is, In a complete directed graph, is it always possible to find a path that includes every vertex exactly once?

One can investigate this question by drawing complete directed graphs with few vertices. You should always be able to find a path that includes every vertex exactly once.

To prove that such a path always exists, we need to find a systematic way to construct a path. Once again, recursive thinking will help. Consider the graph in figure 21.1. The path 1, 3, 2, 4 includes all the vertices except vertex 5. We want to modify this path to form a path that includes vertex 5. Vertex 5 cannot be placed in the path before vertex 1 because the edge from 1 is directed toward 5. Vertex 5 cannot be placed in the path between vertices 1 and 3 because the edge from 3 is directed toward 5. However, vertex 5 can be inserted in the path between vertices 3 and 2 because (3,5) and (5,2) are edges.

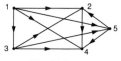

Fig. 21.1

We can generalize the strategy used to add vertex 5 to the path as follows: Assume that you have a complete directed graph with n vertices and have found a path that includes vertices 1, 2, . . . , $n - 1$ exactly once. To add vertex n, find the first vertex, call it m, in the path such that (n, m) is an edge. Insert n just before m in the path. If no such vertex exists, put n at the end of the path. This strategy is the key step in the proof of the following theorem by mathematical induction:

THEOREM. *In a complete directed graph, there is a path that includes every vertex exactly once.*

Proof. We prove the theorem by induction on the number of vertices. If a graph has one vertex, then the path consisting of that vertex satisfies the theorem.

Assume that k is a positive integer and that any complete directed graph with k vertices contains a path that includes every vertex exactly once. Let G be a complete directed graph with $k + 1$ vertices.

Consider the subgraph H of G obtained by deleting one vertex, call it x, and all the edges incident with x. H is a complete directed graph with k vertices, since for each two vertices u, v in H, either (u, v) or (v, u) is an edge in G and thus in H. Therefore the inductive assumption implies that H contains a path x_1, x_2, . . ., x_k that includes every vertex exactly once.

Now consider this path and x in G. If all edges between x and the other vertices are directed toward x, then x_1, . . . , x_k, x is a path in G that includes every vertex exactly once. Otherwise let x_i be the first vertex in the path such that (x, x_i) is an edge. It follows from the choice of x_i and the completeness of G that (x_{i-1}, x) must also be an edge. Hence we can insert x between x_{i-1} and x_i to obtain a path in G that includes every vertex exactly once. It follows by the principle of mathematical induction that the theorem is true for all positive integers.

Returning to round robin tournaments, we can conclude from the previous theorem that a ranking always exists in which the player ranked ith defeated the player ranked $(i + 1)$st. Since there may be more than one path in a complete directed graph that includes every vertex exactly once, this ranking may not be unique.

As the next example illustrates, the variable in a proof by mathematical induction may be the number of edges instead of the number of vertices.

Example 4. A route map for an airline includes n cities, and it is possible to fly from any one of the n cities to any other (not necessarily on a non-stop flight). To save money, the airline wants to eliminate some flights. Is it always possible to eliminate flights in such a way that there remains exactly one route between each pair of cities?

Solution. The route map of the airline can be modeled by a graph in which the vertices represent the cities, and two vertices are connected by an edge if there is a direct flight between the two cities. This graph will be connected, since we assume that it is possible to fly between any two cities. A subgraph of this graph that includes all the vertices and exactly one path between each pair of vertices would represent a route map in which there is exactly one route between each pair of cities.

A subgraph of an undirected graph that is a tree and contains all vertices of the graph is called a *spanning tree*. In a tree, there is precisely one path between each pair of vertices. Hence in graph theory terms, our question is, Does a connected graph always have a spanning tree?

To investigate this question, draw a few examples of connected graphs. You always should be able to find a spanning tree.

To prove that a spanning tree always exists, we need a systematic way of constructing a tree. A connected graph that contains no circuits is a spanning tree of itself. If a connected graph G contains a circuit x_1, x_2, . . ., x_n, x_1, then removing edge $\{x_1, x_2\}$ from the circuit leaves a connected graph (since if x_1, x_2 is included in a path between two vertices, it can be replaced by x_1, x_n, x_{n-1}, . . ., x_2). If this subgraph contains circuits, another edge can be removed. Continuing this process until no circuits remain yields a subgraph that is connected and includes all the vertices of G. Hence this subgraph is a spanning tree of G. We can formalize this process by using mathematical induction.

THEOREM. *A connected graph has a spanning tree.*

Proof. We establish the result by using mathematical induction on the number of edges. If the graph contains one edge, it is its own spanning tree. Thus the theorem is true for a graph with one edge.

Assume that k is a positive integer and that any connected graph with k edges has a spanning tree. Let G be a connected graph with $k + 1$ edges. If G has no circuits, then it is a spanning tree of itself. If G contains a circuit, remove one edge from this circuit. The resulting graph G' is still connected and has k edges. Hence by the inductive assumption, G' has a spanning tree. Since this tree includes all the vertices of G, it is also a spanning tree for G. It follows by the principle of mathematical induction that the theorem is true for all positive integers.

Returning to the route map problem, we can conclude from the theorem that flights can be eliminated in such a way that there remains exactly one route between each pair of cities. Since a graph may have more than one spanning tree, there may be more than one way to eliminate flights.

Additional Problems

The following problems can be presented in a manner similar to the examples above. Answers are in brackets following the question.

1. A binary sequence is a list of 0's and 1's. How many n-digit binary sequences contain an even number of 1's? [2^{n-1}]

2. If n lines in a plane intersect at the same point, how many regions are formed? [$2n$]

3. Suppose there are $2n$ people in a room. What is the maximum number of pairs of people that can be acquainted if there are not three people who know one another? [n^2]

BIBLIOGRAPHY

Crawford, John, and Calvin Long. "Guessing, Mathematical Induction, and a Remarkable Fibonacci Result." *Mathematics Teacher* 72 (November 1979): 613–16.

National Council of Teachers of Mathematics. *Curriculum and Evaluation Standards for School Mathematics.* Reston, Va.: The Council, 1989.

Townsend, Michael. *Discrete Mathematics: Applied Combinatorics and Graph Theory.* Menlo Park: Benjamin-Cummings, 1987.

Tucker, Alan. *Applied Combinatorics.* 2d ed. New York: John Wiley & Sons, 1984.

22

Making Connections through Iteration

David N. Bannard

MATHEMATICAL connections is an area that is stressed in the *Curriculum and Evaluation Standards for School Mathematics* (NCTM 1989). The problem-solving scenario outlined in this chapter evolved from what appeared to be a simple exercise on composition of functions that had the potential for connecting the function concept with that of iteration. The exercise began as follows:

Suppose $g(x) = \dfrac{x + 3}{2}$. Evaluate $g(g(1))$ and $g(g(g(1)))$.

This exercise causes little difficulty for a student who understands function notation. The solution can be expressed as follows:

$$g(1) = 2$$
$$g(g(1)) = g(2) = 2.5$$
$$g(g(g(1))) = g(g(2)) = g(2.5) = 2.75$$

The more interesting question to pursue is what happens if we continue the process to find, for example, $g(g(g(g(1))))$, and so on. This is quite easy to do on a calculator, since the result of each evaluation of the function is plugged back into the function. The key sequence on a standard calculator is 1 $\boxed{+}$ 3 $\boxed{=}$ $\boxed{\div}$ 2 $\boxed{=}$; then take what is in the display (2) and repeat the sequence $\boxed{+}$ 3 $\boxed{=}$ $\boxed{\div}$ 2 $\boxed{=}$ to obtain the next iteration in the sequence, and so on. On a spreadsheet you can obtain the result even faster by defining the first cell as 1, the next cell as the previous cell plus 3 divided by 2, and then copying that formula down through as many cells as you wish. My class produced the spreadsheet in figure 22.1. It was quite obvious to everyone that the sequence of numbers produced was approaching 3. We then investigated the consequences of using different numbers to start the sequence. The sequences in figure 22.2 were obtained.

We continued in this vein. When we started the sequence with 1 000 000, we found that we had to iterate a few more times, but the result always

	A
1	1
2	=(A1 + 3)/2
3	=(A2 + 3)/2
4	=(A3 + 3)/2
5	=(A4 + 3)/2
6	=(A5 + 3)/2
7	=(A6 + 3)/2
8	=(A7 + 3)/2
9	=(A8 + 3)/2
10	=(A9 + 3)/2
11	=(A10 + 3)/2
12	=(A11 + 3)/2
13	=(A12 + 3)/2
14	=(A13 + 3)/2
15	=(A14 + 3)/2

	A
1	1
2	2
3	2.5
4	2.75
5	2.875
6	2.9375
7	2.96875
8	2.984375
9	2.9921875
10	2.99609375
11	2.99804688
12	2.99902344
13	2.99951172
14	2.99975586
15	2.99987793

Fig. 22.1

	A
1	2
2	2.5
3	2.75
4	2.875
5	2.9375
6	2.96875
7	2.984375
8	2.9921875
9	2.9960938
10	2.9980469
11	2.9990234
12	2.9995117
13	2.9997559
14	2.9998779
15	2.999939

	A
1	3
2	3
3	3
4	3
5	3
6	3
7	3
8	3
9	3
10	3
11	3
12	3
13	3
14	3
15	3

	A
1	10
2	6.5
3	4.75
4	3.875
5	3.4375
6	3.21875
7	3.109375
8	3.0546875
9	3.0273438
10	3.0136719
11	3.0068359
12	3.003418
13	3.001709
14	3.0008545
15	3.0004272

	A
1	100
2	51.5
3	27.25
4	15.125
5	9.0625
6	6.03125
7	4.515625
8	3.7578125
9	3.3789063
10	3.1894531
11	3.0947266
12	3.0473633
13	3.0236816
14	3.0118408
15	3.0059204

	A
1	− 10
2	− 3.5
3	− 0.25
4	1.375
5	2.1875
6	2.59375
7	2.796875
8	2.8984375
9	2.9492188
10	2.9746094
11	2.9873047
12	2.9936523
13	2.9968262
14	2.9984131
15	2.9992065

Fig. 22.2

seemed to approach the same number, 3. A natural question arose: Why does the sequence always seem to approach 3? To answer this question, it seemed appropriate to try to generalize or to investigate $g(g(x))$, $g(g(g(x)))$, $g(g(g(g(x))))$, and so on. This became the homework assignment.

A NUMERICAL PERSPECTIVE

Most of my students came back with the following results, although the organization, important for noting trends, was not as clear as written here:

1st iteration: $g(x) = \dfrac{x+3}{2}$

2d iteration: $g(g(x)) = \dfrac{\dfrac{x+3}{2}+3}{2} = \dfrac{x+3+6}{4} = \dfrac{x+9}{4}$

3d iteration: $g(g(g(x))) = \dfrac{\dfrac{x+9}{4}+3}{2} = \dfrac{x+9+12}{8} = \dfrac{x+21}{8}$

4th iteration: $g(g(g(g(x)))) = \dfrac{\dfrac{x+21}{8}+3}{2} = \dfrac{x+21+24}{16} = \dfrac{x+45}{16}$

5th iteration: $g(g(g(g(g(x))))) = \dfrac{\dfrac{x+45}{16}+3}{2} = \dfrac{x+45+48}{32} = \dfrac{x+93}{32}$

The task now was to search for patterns and hope to derive a closed-form formula that would allow us to find the value in a cell based on n, the number of times that we had iterated the function. Certain patterns were obvious to students. The number in the denominator was being doubled with each iteration and thus was forming a geometric sequence in which the nth term was 2^n, where n was the number of times the function was iterated. The sequence 3, 9, 21, 45, 93, . . ., in the numerator was clearly a more complicated pattern than that in the denominator. One of my students observed that the number in the numerator was three times the denominator minus three. However, most students neither noticed this connection nor thought that they would have seen it given more time. Likewise, the student who did see the solution acknowledged that without the denominator, he probably could not have come up with the answer. We now had a formula for the numerator, $x + 3 * 2^n - 3$, but we all agreed that we would like another way to obtain it.

We took a closer look at the fractions above, observing that the numerators follow the pattern $x + 3$, $x + 3 + 6$, $x + 9 + 12$, $x + 21 + 24$, $x +$

$45 + 48$. Therefore the next term should be $x + 93 + 96$, or $x + 189$. Of course, this can and should be verified by considering the sequence $3, 9, 21, 45, 93$ and computing the differences between successive terms, $6, 12, 24, 48$.

Returning to the algebra used in iterating $g(x)$, we reexamined our work as follows, this time using a slightly more detailed view:

$$g(x) = \frac{x + 3}{2}$$

$$g(g(x)) = \frac{\dfrac{x + 3}{2} + 3}{2} = \frac{x + 3 + 6}{4} = \frac{x + 9}{4}$$

$$g(g(g(x))) = \frac{\dfrac{x + 3 + 6}{4} + 3}{2} = \frac{x + 3 + 6 + 12}{8} = \frac{x + 21}{8}$$

$$g(g(g(g(x)))) = \frac{\dfrac{x + 3 + 6 + 12}{8} + 3}{2} = \frac{x + 3 + 6 + 12 + 24}{16} = \frac{x + 45}{16}$$

$$g(g(g(g(g(x))))) = \frac{\dfrac{x + 3 + 6 + 12 + 24}{16} + 3}{2}$$

$$= \frac{x + 3 + 6 + 12 + 24 + 48}{32} = \frac{x + 93}{32}$$

A summary of the algebra above is given in table 22.1.

TABLE 22.1

Iteration (n)	Numerator	Denominator
1	$x + 3$	2
2	$x + 3 + 6$	4
3	$x + 3 + 6 + 12$	8
4	$x + 3 + 6 + 12 + 24$	16
5	$x + 3 + 6 + 12 + 24 + 48$	32
etc.		

The numerator is x plus a geometric series. Recalling that the sum of n terms of a geometric series is given by $S(n) = a\left(\dfrac{1 - r^n}{1 - r}\right)$, where a is the first term and r is the constant ratio, we obtained $S(n) = 3\left(\dfrac{1 - 2^n}{1 - 2}\right)$, which simplifies to $3 \cdot 2^n - 3$.

It was now possible to examine the eventual formula, $\dfrac{x + 3 \cdot 2^n - 3}{2^n}$.

This simplifies to $\dfrac{x}{2^n} + \dfrac{3 \cdot 2^n}{2^n} - \dfrac{3}{2^n} = 3 + \dfrac{x}{2^n} - \dfrac{3}{2^n}$. An interesting discussion of the growth of the function 2^n followed. It then became clear why the iterated function values were converging to 3, regardless of the initial value of x we chose.

As we progressed in our investigation, students developed a greater appreciation for the connection among functions, iteration, sequences, and series and for the supporting interplay between reasoning and algebraic skills. However, questions still remained. Do other functions also converge, and if so, to what number do they converge? Do all linear functions converge, and if not, which do and which do not? And to what? These questions became the next assignment for my class. We discussed that if $g(x) = \dfrac{x + h}{k}$, the iterated sequence obtained by evaluating $g(g(x))$, $g(g(g(x)))$, and so on, will converge for $|k| > 1$ and diverge for $|k| \le 1$. Equivalently, if $g(x)$ is defined in the somewhat more familiar form of a linear function, $g(x) = mx + b$, the iterated sequence will converge for $|m| < 1$ and diverge for $|m| \ge 1$. By examining the convergence for different values of h and k, some of my students were able to find that the number to which these sequences converge is $\dfrac{h}{k - 1}$ or where $\dfrac{x + h}{k} = x$. This led to another technique for looking at the question—graphical analysis.

A GRAPHICAL PERSPECTIVE

The graph in figure 22.3 illustrates one way of analyzing the situation. To begin, graph $y = g(x)$, a linear function whose slope is less than 1 for $k > 1$, and then graph $y = x$. Begin with a value of x, say 10. To find $g(10)$, move vertically up to the line $g(x) = \dfrac{x + 3}{2}$. Since the y-coordinate of the resulting point of intersection will become the next x-coordinate, move horizontally to the line $y = x$. At this point y and x are equal, so this is our next x-value. Go vertically down to intersect the line $y = g(x)$, then horizontally to the line $y = x$, vertically to the line $y = g(x)$, and so on. This process converges on the point where the lines $y = g(x)$ and $y = x$ intersect. If this method is used with any linear function $y = g(x)$ whose slope is between 0 and 1, the iterated function can be seen to converge to the intersection of $y = g(x)$ and $y = x$. The method can also be used to show that the iterated linear function $g(x)$ diverges when the slope is greater than 1. It is left to the reader to examine the behavior of linear functions whose slopes are less than 0. They should be examined, both by iterating them using a spreadsheet or calculator and by using graphical analysis.

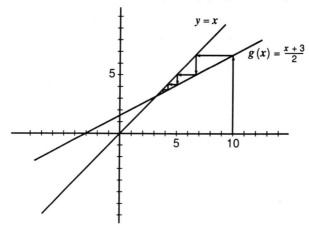

Fig. 22.3

Finally, the reader is encouraged to examine other functions to see under what conditions the corresponding sequences converge or diverge. Some interesting functions to look at are listed below.

$$f(x) = \sqrt{x}$$
$$f(x) = \sin(x)$$
$$f(x) = \cos(x)$$

$$f(x) = x^2 - 2$$
$$f(x) = 1 - x^2 \text{ for } -2 \le x \le 2$$
$$f(x) = \frac{1}{2}(x - 2)(3x + 1)$$

SUMMARY

My students were highly motivated throughout our study of this problem primarily because of our ability to experiment and discover new aspects of the situation together. Only a calculator was needed for the investigation. However, a spreadsheet or a simple computer program makes the investigation even more dramatic and fun. The investigation was open-ended. It was exciting for the students because they found they were discovering patterns they had not expected to find. As a result, they were motivated to learn the mathematics necessary to continue the investigation. It also set the stage for introducing students to a new field of mathematics called "chaos." Classroom possibilities for topics from this field are the focus of the next chapter.

REFERENCE

National Council of Teachers of Mathematics. *Curriculum and Evaluation Standards for School Mathematics.* Reston, Va.: The Council, 1989.

23

Putting Chaos into the Classroom

Robert L. Devaney

C HAOS, fractals, dynamics—these terms have been very much in the news in recent years. Spectacular computer-graphics images of the Mandelbrot set have appeared in full color in virtually every popular science magazine. Fractals, especially Julia sets, have appeared in art museums, on postcards, in Hollywood films, and elsewhere. And James Gleick's book *Chaos: Making a New Science* (1987) remained on the *New York Times* best-seller list for over six months.

Why all the interest in these fields? There are many reasons for this. One is the great beauty and intricacy of the computer-graphics images of such mathematical objects as the Mandelbrot set and Julia sets (Mandelbrot 1982; Peitgen and Richter 1986). See the following pages for examples. A second reason is that these concepts give scientists in all disciplines new and exciting tools with which to attack formerly intractable problems. A third reason, one that has not yet been thoroughly appreciated, is that many of these topics are quite accessible. Students, even high school students, with access to computers and computer graphics can appreciate the basic concepts of chaos, fractals, and dynamics. More important, using the computer as an experimental tool, these students can quickly reach the forefront of knowledge in these fields—even simple quadratic functions of a real variable are not completely understood when they are treated as a dynamical system. To a high school or beginning college student who has never encountered anything but centuries-old mathematics (and probably regards mathematics as a "dead" subject), this realization comes as quite a shock. When coupled with the natural beauty of this subject, this knowledge gives students a very different view of mathematics as a scientific discipline.

The goal of this chapter is to give a few examples of how these topics can be incorporated into mathematics courses at the precalculus level and beyond. It is our experience that these topics go over very well with groups of motivated high school students. Many of these topics offer a way to reinforce students' grasp of such precalculus topics as the quadratic formula, graphing, trigonometric functions, and so forth. Consequently, these topics make

Julia set of a complex sine function

The Mandelbrot set and several magnifications

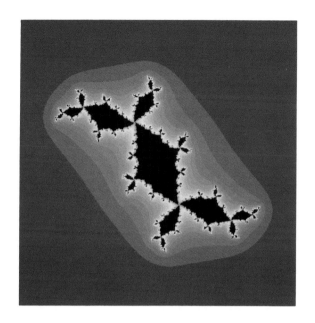

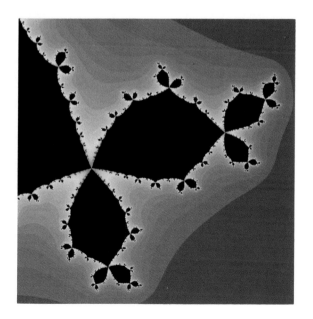

The fractal rabbit:
Julia set of $z^2 - 0.15 + 0.75i$ and a magnification

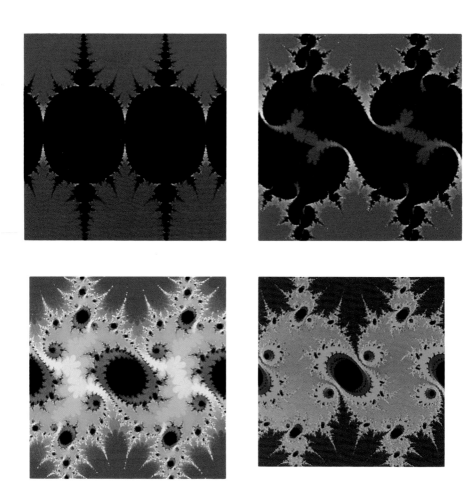

The exploding Julia set for the complex sine function

a perfect mathematics course for second-semester high school seniors. Rather than give students a diluted survey of a few calculus topics, it might be better to both entice the students and prepare them for later mathematics study by teaching a few topics in dynamics instead. These topics would also be a natural course for liberal arts majors who are required to take a course that gives them a "mathematical experience."

DYNAMICAL SYSTEMS

The terms *chaos* and *fractal* are topics that arise quite naturally in the field of mathematics known as dynamical systems. Basically, a dynamical system is any process that moves or changes in time. Examples of dynamical systems abound in all the natural and social sciences. Simple examples are the changing weather, the fluctuations of the Dow-Jones average, simple chemical reactions, and the motion of the planets. You can undoubtedly think of many more examples in virtually every field of science. Indeed, the *omnipresence* of dynamical systems is one of the reasons this field of mathematics has received so much attention recently.

The major question in dynamics is whether one can predict the eventual or future behavior of the system as it moves or changes in time. From our examples above, it seems that sometimes the answer is yes, and sometimes it is no. For example, prediction of the weather a week from now or the Dow-Jones average a day from now seems all but impossible. By contrast, planetary motion and simple chemical reactions are quite predictable—we know that the sun will rise tomorrow morning and that when we add cream to our coffee, the resulting chemical reaction will not be an explosion.

From these examples, it may seem that the reason certain dynamical systems are unpredictable is that they depend on a large number of variables. This is certainly true for many meteorological and economic systems. However, one of the fundamental discoveries in recent years is that systems that depend on very few variables—sometimes only one variable—may behave just as unpredictably as the weather or the Dow-Jones average. This fact is easy to demonstrate using simple mathematical dynamical systems.

ITERATION

One of the simplest types of dynamical systems is the iterated function. This concept is easily explained to any student who has access to a scientific calculator. Given one of the functions represented by a key on the calculator and an initial input x_0, the basic problem in dynamics is to compute the successive iterates of the function applied to x_0 and then predict what will eventually happen. For example, consider the square-root function $S(x) = \sqrt{x}$. Given any initial input, say $x_0 = 256$, one may compute successively

$$x_0 = 256,$$
$$x_1 = \sqrt{256} = 16,$$
$$x_2 = \sqrt{16} = 4,$$
$$x_3 = \sqrt{4} = 2,$$
$$x_4 = \sqrt{2} = 1.41 \ldots ,$$

and so forth. The resulting list of numbers, $x_0, x_1, x_2, \ldots$ is called the orbit of x_0 under S; note that these numbers change in time; so this is an example of a dynamical system. The basic question is, as before, can we predict what will happen to the orbit? Obviously, for the square-root function, predicting the behavior of any orbit (for positive x_0) is easy: all orbits tend to 1.

It is easy to illustrate iteration and orbits using a calculator. Simply enter any number into the calculator and repeatedly strike the square-root key: the resulting list of numbers is the orbit of the initial input.

Another easy example is the squaring function $T(x) = x^2$. Clearly, if $|x_0| > 1$, the orbit of x_0 tends to infinity, whereas if $|x_0| < 1$, the orbit tends to 0. The orbit of 1 is special; 1 is a *fixed point* because it remains fixed for all iterations. Similarly, -1 is an *eventually fixed point*, since $T(-1) = 1$.

Let us introduce some notation for iteration. Given a function F, let us denote by F^n the nth iterate of F. That is,

$$F^2(x) = F(F(x)),$$
$$F^3(x) = F(F(F(x))),$$

and so forth. It is important to point out that $F^2(x)$ does *not* mean the product $F(x) \cdot F(x)$. Thus, the orbit of x is the infinite list of numbers

$$x, F(x), F^2(x), \ldots , F^n(x), \ldots ,$$

and the basic question for these dynamical systems is, Can we predict the eventual fate of orbits? Here is another iteration. Can you predict the fate of any orbit of $C(x) = \cos x$? The answer to this question comes as quite a surprise to student and teacher alike. If $C(x)$ is computed in radians, then any orbit of C tends eventually to 0.73908. . . . This was probably not your first guess for the fate of any orbit of cosine, but we will see where this number comes from in a moment.

One final example—consider the orbits of x_0 with $0 < x_0 < 1$ under iteration of the simple quadratic function $F(x) = 4x(1 - x)$. Table 23.1 lists several such orbits (computed to only three decimal places). Notice several things. First, you can't compute these orbits very easily with a calculator— you really need a computer or a programmable calculator. (The computer program to find orbits is simple—one loop generates the orbit of this, or any, function.) Second, note the great diversity among the orbits. The orbit of .5 is simple; it eventually becomes fixed at 0 ($F(0) = 0$, so $F^n(0) = 0$ for all n). Similarly, $F(3/4) = 3/4$; so 3/4 is also a fixed point. Finally, look closely at the nearby orbits of 0.51 and 0.749. After a very few iterations,

TABLE 23.1

VARIOUS ORBITS OF $4x(1 - x)$

$x_0 =$	0.1	0.25	0.3	0.5	0.51	0.749	0.8
1	0.36	0.75	0.84	1	0.999	0.752	0.640
2	0.922	0.75	0.538	0	0.001	0.746	0.922
3	0.289	0.75	0.994	0	0.006	0.758	0.289
4	0.822	0.75	0.022	0	0.025	0.734	0.822
5	0.585	0.75	0.088	0	0.099	0.781	0.585
6	0.971	0.75	0.321	0	0.356	0.684	0.971
7	0.113	0.75	0.871	0	0.917	0.865	0.113
8	0.402	0.75	0.448	0	0.301	0.466	0.402
9	0.961	0.75	0.989	0	0.842	0.995	0.962
10	0.148	0.75	0.043	0	0.530	0.018	0.148
11	0.504	0.75	0.166	0	0.996	0.071	0.504
12	1.000	0.75	0.554	0	0.014	0.262	1.000
13	0.000	0.75	0.988	0	0.058	0.774	0.000
14	0.001	0.75	0.045	0	0.219	0.699	0.001
15	0.004	0.75	0.174	0	0.686	0.841	0.004
16	0.015	0.75	0.575	0	0.861	0.534	0.015
17	0.060	0.75	0.977	0	0.477	0.995	0.059
18	0.227	0.75	0.089	0	0.998	0.017	0.222
19	0.703	0.75	0.321	0	0.007	0.073	0.690
20	0.836	0.75	0.873	0	0.031	0.272	0.856

these orbits bear no resemblance whatsoever to the orbits of their neighbors 0.5 and 0.75. This simple observation has had profound ramifications in many areas of mathematics and science in recent years. This is *chaos*—nearby points have orbits that behave in vastly different ways.

GRAPHICAL ANALYSIS

Graphical analysis can be used to supplement understanding of iteration. This process allows a student to follow simple orbits of certain functions using only the graph of the function (not its iterates).

The process works like this. Superimpose the graph of the function $y = F(x)$ and the line $y = x$. Intersections of the two graphs give the fixed points of F. For a nonfixed point x_0 we will display its orbit on the diagonal line. Begin at (x_0, x_0) and draw a vertical line to the graph. We reach the point $(x_0, F(x_0))$. Now draw a horizontal line from the graph to the diagonal. We reach the diagonal at $(F(x_0), F(x_0))$. That is, the process of applying F to x_0 is given geometrically by drawing a vertical line to the graph then a horizontal line back to the diagonal. Repeating this procedure starting from $(F(x_0), F(x_0))$ yields the second point on the orbit, namely $(F^2(x_0), F^2(x_0))$.

Figure 23.1 shows how this procedure applies to $S(x) = \sqrt{x}$. Note that all orbits tend to 1, a fact that was verified earlier using the calculator. Figure 23.2 displays the graphical analysis of $C(x) = \cos x$ and, in particular, explains where the number 0.73908 . . . comes from. Figure 23.3 exhibits a chaotic orbit for the quadratic function $F(x) = 4x(1 - x)$.

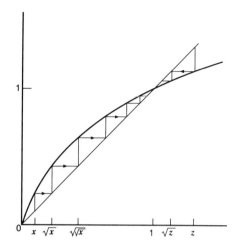

Fig. 23.1. Graphical analysis of $S(x) = \sqrt{x}$

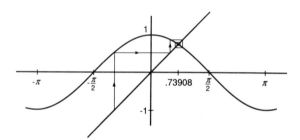

Fig. 23.2. Graphical analysis of $C(x) = \cos x$

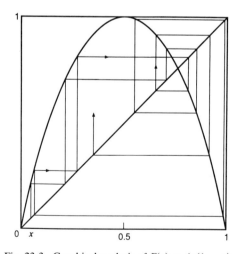

Fig. 23.3. Graphical analysis of $F(x) = 4x(1 - x)$

LOGISTIC FUNCTIONS

The family of functions $F_\mu(x) = \mu x(1 - x)$ is called the *logistic* family. The parameter μ here is usually taken between 0 and 4. This family of functions exhibits a stunning array of complex behavior as the parameter μ increases. Recall that when $\mu = 4$, many chaotic orbits were present. When μ is small, however, the orbits are quite tame. When $\mu \leq 1$, graphical analysis shows that if we restrict x_0 to $0 < x_0 < 1$, then all orbits tend to 0. See figure 23.4a. If $1 \leq \mu \leq 3$ (fig. 23.4b), then all orbits tend to the fixed point given by finding the nonzero root of

$$\mu x(1 - x) = x.$$

Above $\mu = 3$, the dynamics are more complicated. Students have a great deal of fun trying to determine experimentally what happens to a typical orbit in this range using the simple iteration program alluded to above. For example, for μ in the range $3 < \mu < 3.43 \ldots$, it appears that all orbits eventually tend to cycle back and forth between two numbers: this is a period 2 cycle. Then, for μ in the range $3.43 \ldots < \mu < 3.53 \ldots$, it appears that all orbits eventually cycle with period 4. As μ increases, we see smaller and smaller μ-intervals in which we get cyclic behavior of period 2^n.

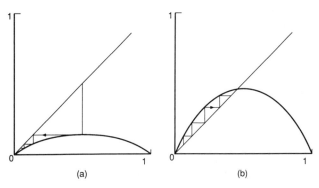

(a)   (b)

Fig. 23.4.

Eventually, though, all we see is chaos. There are, however, a very few μ-values that lead to similar cyclic behavior; for example, $\mu = 3.83$ yields cycling of order 3.

How do we understand all this? The computer (and computer graphics) gives us the answer. In figure 23.5 we have plotted the orbit diagram of the logistic family. This diagram plots the orbit of 1/2 versus μ for 400 μ-values between 3 and 4. The orbit of 1/2 for the corresponding F_μ is plotted vertically. Pseudocode to generate the orbit diagram is given in figure 23.6.

Several comments are in order. First, we have plotted only the last 100 of 150 points on each orbit. This eliminates the "transients" before the orbit

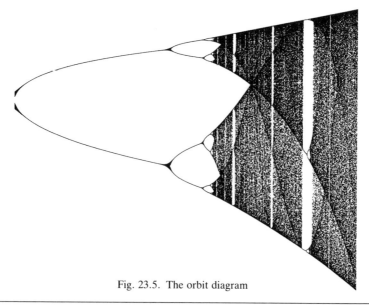

Fig. 23.5. The orbit diagram

Step 0. Initialize $\mu = 3$.
Step 1. Let $x_0 = 0.5$, $i = 1$.
Step 2. Compute $x_i = \mu x_{i-1}(1 - x_{i-1})$.
Step 3. If $i < 50$, increment i and return to step 2. If $50 \leq i \leq 150$, plot (μ, x_i).
          Increment i and return to step 2.
Step 4. If $i = 151$ and $\mu < 4$, then increment μ by 0.0025 and return to step 1.
Step 5. If $\mu \geq 4$, stop.

Fig. 23.6. Pseudocode to produce the orbit diagram of $F_\mu(x) = \mu x(1 - x)$ for 400 equally spaced μ-values between 3 and 4

settles down. Note that the chaotic behavior is clearly visible, as are the cyclic regimes. Amazingly, virtually the same image results no matter which initial input x_0 satisfying $0 < x_0 < 1$ is chosen. Even more amazing is the fact that almost any function that is quadratic-like has the same orbit diagram (subject to stretching or squeezing). Without being too precise, by quadratic-like we simply mean that the family of functions has graphs that, on a certain interval, look like the logistic family. For example, the family $S_\lambda(x) = \lambda \sin x$ with $0 \leq x \leq \pi$ and $0 < \lambda < \pi$ is quadratic-like.

When students see images like this, they marvel at the complexity that such a simple iteration exhibits. When they hear that much of the behavior that they see in the orbit diagram was observed for the first time a mere fifteen years ago, they become quite intrigued. Finally, when they find out that this family—a simple, quadratic function of a real variable—is not completely understood, they begin to realize that mathematical research is indeed an interesting and exciting field.

AN APPLICATION

Students rarely ask for applications of iteration. However, there are many, and the number of applications will undoubtedly increase in the future as the computer, with its ability to compute thousands of points on an orbit quickly, becomes even more important. Newton's method for finding roots is one natural example as are all the numerical methods for solving differential equations. Mortgage and interest rate calculations also involve iteration. Here is an application of a different sort.

Simple difference equations lead to iteration problems; the logistic equation from mathematical biology is one example. To introduce this equation, suppose that a single species is living and reproducing in a controlled laboratory environment so that the population may be computed at each generation. The biologist creates and uses mathematical models to predict the ultimate destiny of this population. It is important that the biologist be able to predict whether the species will become extinct, whether the population will vary cyclically, and so forth.

Rather than work with the actual numbers of the species, let us work with the percentage of a predetermined limiting population. That is, let P_n denote the percentage of the limiting number of the species alive during generation n. So $P_n = 0$ means that the species has become extinct, whereas P_n near 1 indicates serious overcrowding.

One of the simplest models for this situation is the logistic equation

$$P_{n+1} = \mu P_n(1 - P_n)$$

where, as before, μ is a parameter that now may be interpreted as depending on ecological conditions (availability of food, climate, etc.). Given an initial population P_0, the biologist can use the logistic equation to predict future populations. Indeed,

$$P_1 = \mu P_0(1 - P_0) = F_\mu(P_0)$$
$$P_2 = \mu P_1(1 - P_1) = F_\mu^2(P_0)$$

and, in general,

$$P_n = F_\mu^n(P_0).$$

So we see that computing the orbit of an initial population P_0 satisfying $0 < P_0 < 1$ under the logistic function is precisely the solution to the biologist's problem. Thus, as we observed above, for μ-values below 1, the population becomes extinct. For $1 < \mu \leq 3$, the population tends to reach an equilibrium state. Above $\mu = 3$, the population may vary cyclically or chaotically.

Thus simple biological models can lead to both cyclic and chaotic behavior, a fact that has only recently come to be appreciated by biologists.

JULIA SETS

By far the most intriguing dynamical systems arise from iterations in the complex plane, not on the real line as in previous sections. Many students encounter complex numbers in advanced algebra. Learning complex arithmetic with an eye toward iterating complex dynamical systems is an ideal way to combine theoretical mathematics and computer experimentation.

One of the most beautiful areas of contemporary research in mathematics concerns the iteration of complex functions of the form

$$F(z) = z^2 + c.$$

Here, both z and c are complex numbers, $z = x + iy$ and $c = c_1 + ic_2$. That is, if $x_1 + iy_1 = F(z)$, then complex arithmetic yields

$$x_1 = x^2 - y^2 + c_1$$
$$y_1 = 2xy + c_2.$$

These formulas are employed over and over during the iteration.

Of interest in complex iteration is the *filled-in Julia set*. This is the set of points whose orbits do not tend to infinity. As an example, consider the complex squaring function $T(z) = z^2$. Let us denote the distance of a complex number z from the origin by $|z|$ (this is commonly called the modulus of z). If $z = x + iy$, then

$$|z| = \sqrt{x^2 + y^2}.$$

For the squaring function, we compute easily

$$|T(z)| = |T(x + iy)| = \sqrt{(x^2 - y^2)^2 + (2xy)^2}$$
$$= \sqrt{x^4 + y^4 + 2x^2y^2}$$
$$= \sqrt{(x^2 + y^2)^2}$$
$$= |z|^2.$$

Therefore, if $|z| > 1$, then $|T(z)| > |z|$. By contrast, if $0 < |z| < 1$, then $|T(z)| < |z|$. If $|z| = 1$, then $|T(z)| = 1$, and the orbit remains forever on the unit circle. This means that the complex squaring function behaves in essentially the same manner as the real squaring function: if $|z| < 1$, then the orbit of z tends toward 0, whereas if $|z| > 1$, the orbit of z tends to infinity. The filled-in Julia set (the orbits that don't escape to infinity) is precisely the set of points with $|z| \leq 1$, that is, the unit disk.

This filled-in Julia set is not tremendously exciting, but the Julia sets for $F(z) = z^2 + c$ with nonzero c are almost always geometrically interesting. Their boundaries are known as fractals. Figure 23.7 shows a variety of these filled-in Julia sets. Note the changes in shape as the parameter c varies.

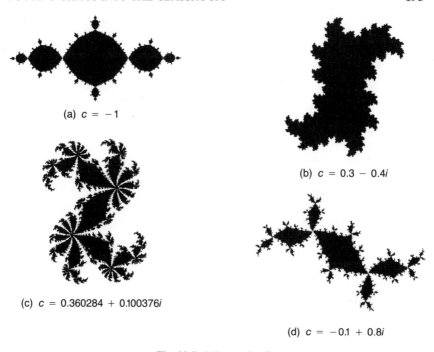

(a) $c = -1$

(b) $c = 0.3 - 0.4i$

(c) $c = 0.360284 + 0.100376i$

(d) $c = -0.1 + 0.8i$

Fig. 23.7. Julia sets for Q_c

The relation of these images to the chaotic behavior noted above is delicate. Any point colored white in figure 23.7 has orbit that tends to infinity, whereas any orbit corresponding to a point in the interior of the black region tends to a periodic cycle. The points that lie on the boundary of the black region have chaotic orbits. This is true because nearby orbits tend far away—neighboring points that are white have orbits that behave quite differently from neighboring points that are black. The boundary of the black region is called the Julia set, and it is important for this reason: it is the set on which all the chaos occurs.

The algorithm to produce the Julia sets of $F(z) = z^2 + c$ is easy to explain. Let us assume that $|c| < 2$. All the interesting pictures occur for these small c-values. If any point has an orbit that eventually enters the region beyond the circle of radius 2 centered at the origin, then that orbit necessarily escapes to infinity. Hence we need only iterate until the orbit leaves this circle to color a point white. If no point on the orbit escapes during the first twenty iterations, we agree to color that point black. Figure 23.8 displays pseudocode that can create these images.

Images such as those displayed in figures 23.5 or 23.7 are easily produced on today's computers. They can even be displayed (in low resolution) on programmable graphing calculators!

There is, of course, much more to the story of Julia sets and their dynam-

Step 0. Input CREAL, CIMAG.
Step 1. Initialize $x_0 = -2$, $y_0 = -2$.
Step 2. Set $j = 1$.
Step 3. Compute:

$$x_j = x_{j-1}^2 - y_{j-1}^2 + \text{CREAL}$$
$$y_j = 2x_{j-1}y_{j-1} + \text{CIMAG}$$
$$r_j = x_j^2 + y_j^2$$

Step 4. Test:
   A. If $r_j < 4$ and $j \le 20$, then increment j and return to step 3.
   B. If $r_j < 4$ but $j = 21$, then plot (x_0, y_0).
   C. If $r_j \ge 4$, go on to step 5.
Step 5. If $y_0 < 2$, replace y_0 by $y_0 + .01$. Return to step 2.
Step 6. If $y_0 \ge 2$, then let $y_0 = -2$ and replace x_0 by $x_0 + .01$. Return
   to step 2.
Step 7. If $x_0 < 2$, return to step 2.
Step 8. Stop.

Fig. 23.8. Pseudocode to produce Julia sets of $F(z) = z^2 + c$ where c is the complex
number CREAL + iCIMAG

ics. This short section is merely intended to show how computer graphics
and simple iteration may be combined to yield fascinating insights into
problems and ideas in contemporary mathematics.

CONCLUSION

Topics in dynamical-systems theory offer ideal opportunities for teachers
to expose their students to exciting, alluring, and contemporary ideas in
mathematics. These topics need the use of the computer as an experimental
tool, and when combined with computer graphics, they give the student a
very different and exhilarating mathematical experience.

BIBLIOGRAPHY

Devaney, Robert L. "Fractal Patterns Arising in Dynamical Systems." In *The Science of Fractal Images*. New York: Springer-Verlag, 1988.

———. *Chaos, Fractals, and Dynamics: Computer Experiments in Mathematics*. Menlo Park, Calif.: Addison-Wesley Publishing Co., 1990.

Gleick, James. *Chaos: Making a New Science*. New York: Viking Penguin, 1987.

Mandelbrot, Benoit. *The Fractal Geometry of Nature*. San Francisco: W.H. Freeman & Co., 1982.

Peitgen, Heinz-Otto, and Peter Richter. *The Beauty of Fractals*. Heidelberg: Springer-Verlag, 1986.

Algorithms: You Cannot Do Discrete Mathematics without Them

Stephen B. Maurer
Anthony Ralston

THERE have always been algorithms in secondary school mathematics. To solve a quadratic equation by formula or to multiply two linear polynomials by FOIL is to carry out an algorithm, albeit with an emphasis on rote performance of a procedure rather than an emphasis on analyzing or discovering procedures. But very little attention has been given to devising alternative algorithms for a task and then comparing them for ease and speed of use. Finally, algorithms in school mathematics have always been applied to small problems—those tractable by pencil and paper—and so there has been no need either to state or to follow the algorithms precisely. Shortcuts and variations suitable for hand computation have sometimes been encouraged.

Changes are needed. Today almost all the myriad applications of mathematics are performed by calculators or computers. Furthermore, the computer and calculator power to do large problems is available in schools. It is no longer necessary to give students "toy" problems, such as polynomials of degree no greater than four and linear systems with, at most, three variables. (In business and science, systems with tens of thousands of variables are solved.) Nor is it necessary to perpetuate the fiction that coefficients are always integers or simple rational numbers. However, to solve large problems involving messy numbers, you *must* be systematic, that is, truly algorithmic (Maurer 1984). And you must be concerned about the efficiency of algorithms because, no matter how fast your computer is, you can be sure that someone will soon come along with a problem large enough to tax it.

Not only does the algorithmic approach allow the use of realistic problems, it also forces you to generalize, that is, to consider all aspects of a problem, not just special cases. (Of course, special cases can—and should—be used to prompt generalizations.) Being forced to generalize may not sound like an advantage, but mathematical knowledge is only really useful when it allows a class of problems rather than specific cases to be understood and attacked. As we shall see, writing general algorithms requires more attention to mathematical notation than has been typical in secondary school.

For generalization through the use of algorithms to be effective, you need problem-solving methods attuned to this approach. Two such methods are the *recursive paradigm* (RP) and the *inductive paradigm* (IP). The RP embodies an approach in which we solve a problem by supposing we already know how to solve it for a simpler case than the one at hand. Then we reduce the problem at hand to the simpler, known one and, voilà, the general solution is at hand. The IP is a powerful tool when an approach to a solution is not obvious but where some computation may enable you to discern a pattern that will lead you to the solution. The paradigm is compute, conjecture, prove (that the conjecture is correct). In this chapter we shall apply these two paradigms to a variety of problems.

Still another reason to use an algorithmic approach to teaching is that the variety of calculation tools now available in schools—pencil and paper, calculator, computer—means that what is a good procedure (i.e., algorithm) in one medium may not be so in another. Only by getting students to think procedurally (i.e., algorithmically) can this point be taught effectively and can students develop the habit of determining which algorithms are best for which devices.

It is often said that the best way to learn a subject is to teach it. A contemporary paraphrase of this is that if you can teach something to a computer (i.e., program it), then you must really understand it. Getting students to derive algorithms is essentially the same intellectual task as programming but without the niggling details.

One more introductory remark—we haven't mentioned the word *discrete* yet, except in our title. In effect, *any* mathematics where answers must be computed involves algorithms; for instance, to find zeros of most continuous functions requires approximation algorithms. To keep within the scope of this volume, we will stick to topics where algorithms (*finite* procedures) can give *exact* solutions. In this sense, a topic like the multiplication of polynomials is part of discrete mathematics.

AN ALGORITHMIC APPROACH TO STANDARD TOPICS

This section illustrates the value of an algorithmic approach to three standard discrete topics in secondary school mathematics. For each topic we will present the *outcome* of an algorithmic approach. In class you would want to lead up to these outcomes less abruptly than we do; in later sections we indicate some ways to do this.

Polynomial Evaluation

Evaluating polynomials is a staple of the secondary school mathematics curriculum. How should it be done? Consider

$$p(x) = 5x^4 - 2x^3 + 3x^2 - 7x + 4. \qquad (1)$$

The traditional brute-force method entails substituting a value for x into the terms in (1) in the order shown. For high-degree polynomials (which, for example, occur when solving differential equations using power series and in algebraic coding theory), this is very tedious. Even for low-degree polynomials this is not particularly efficient when a calculator is used. Whether the polynomial is of high or low degree, there is a better way than the brute-force method. The better way appears in most algebra 2 texts, although usually rather cursorily, and is called *synthetic division* or *Horner's method.* To evaluate the polynomial above at $x = 3$, synthetic division uses the tableau

5	-2	3	-7	4	$\underline{3}$
	15	39	126	357	
5	13	42	119	361	

to arrive at $p(3) = 361$. Why is this method better? Because calculating the tableau above requires only four multiplications whereas the brute-force method requires seven multiplications (three to compute x^4 and four for the coefficient multiplications) or even more if you don't save each power of x as you compute it.

Synthetic division is certainly an algorithm, but how would you describe it as a procedure that works for any degree polynomial and that could become the basis of a computer program? The key point is that synthetic division embodies rewriting (1) in *nested* form as

$$p(x) = x(x(x(5x - 2) + 3) - 7) + 4, \qquad (2)$$

and then evaluating this equation inside out. For example, let $x = 3$, calculate 5×3, subtract 2, multiply by 3, and so on. To describe this idea mathematically, first write a general polynomial of degree n as

$$p_n(x) = a_n x^n + a_{n-1} x^{n-1} + \ldots + a_1 x + a_0, \qquad (3)$$

and then write (3) in nested form as

$$p_n(x) = \underbrace{x(x(\ldots x}_{(n-1)x\text{'s}} (a_n x + a_{n-1}) + a_{n-2}) + \ldots + a_1) + a_0. \qquad (4)$$

Using this notation rather than just illustrating by example, as above, requires more initial effort by the student, but it is worthwhile. The ability to understand and deploy mathematical notation should be one of the most important results of secondary school mathematics. For polynomials, in particular, the general notation in (3) is valuable; we shall make crucial use of it in the next example.

The form in (4) suggests evaluating the polynomial from inside out using the algorithm in figure 24.1. Students who can understand this algorithm will have a deeper understanding of Horner's method than can be obtained by any number of specific examples.

Input n	[Degree of polynomial]
x	[Value of variable]
a_i, i = 0, . . ., n	[Polynomial coefficients]
Algorithm Horner	
$p \leftarrow a_n$	[Assign a_n to p]
for j = n−1 **down to** 0	
$p \leftarrow xp + a_j$	
endfor	
Output p	[Value of polynomial at x]

Fig. 24.1. Algorithm for polynomial evaluation

Polynomial Multiplication

The standard approaches to multiplying polynomials in secondary school are to use the FOIL technique (First, Outside, Inside, Last) for multiplying linear polynomials in algebra 1 and then to use the generalized distributive approach (multiply all terms of one polynomial by each term of the other and collect like powers) in algebra 2. This approach may be satisfactory for the highest-degree polynomials one is likely to want to multiply by hand, but it gives little insight into how to describe the multiplication of two general polynomials to a computer. Since the latter task is much more important mathematically than hand polynomial multiplication, it is worthwhile to develop the general algorithm.

The generalized distributive-law approach provides the key observation for an algorithmic approach to polynomial multiplication. The coefficient of x^i in the product of two polynomials consists of the sum of all products of coefficients, one from each polynomial, whose subscripts sum to i. This observation will not be obvious to many students; in a later section we illustrate how they may be led to conjecture it for themselves. Suppose we want

$$\sum_{k=0}^{m+n} c_k x^k = \left(\sum_{i=0}^{m} a_i x^i\right)\left(\sum_{j=0}^{n} b_j x^j\right). \tag{5}$$

Notice that we have switched to representing polynomials using summation notation. We and, we think, you and your students will find continual writing of (3) very tedious indeed. This is motivation enough for introducing summation notation that, except for subscripts, is the most important piece of mathematical notation not currently emphasized in secondary school that students should be comfortable with (not just familiar with) at the end of secondary school mathematics.

The observation above means that c_k in (5) is given by

$$c_k = \sum_{i=0}^{k} a_i b_{k-i} \qquad k = 0, 1, \ldots, m+n \qquad (6)$$

with the stipulation that $a_i = 0$ if $i > m$, and $b_{k-i} = 0$ if $k-i > n$. This stipulation is necessary because either subscript in the summation could have a value higher than the degree of the corresponding polynomial (e.g., if $m = 3, n = 4$, and $k = 6$, then $a_5 b_1$ is one of the terms in the summation, but $5 > m$).

The idea embodied in (6) is most conveniently represented by an algorithm as in figure 24.2. We don't claim that this method of multiplying polynomials is easier than FOIL or the generalized distributive law. It's not. We do claim, however, that once the method is learned in this form, the student can not only multiply any two polynomials but, more important, the student will understand polynomial multiplication in the sense of being able to apply it with symbolic as well as numerical coefficients.

```
Input    m; a_i, i = 0, ..., m       [Degree and coefficients of one polynomial]
         n; b_j, j = 0, ..., n       [Degree and coefficients of second polynomial]
Algorithm PolyMult
       for k = 0 to m+n
            c_k ← 0                                           [Initialize c_k]
            for i = 0 to k
                 if i ≤ m and k−i ≤ n then c_k ← c_k + a_i b_{k-i}
            endfor
       endfor
Output   c_k, k = 0, 1, ..., m+n      [Coefficients of product polynomial]
```

Fig. 24.2. Algorithm for polynomial multiplication

Gaussian Elimination

Another standard topic in secondary school mathematics is solving simultaneous linear equations. It is rare today for a secondary school student to see—much less deal with—a system of equations whose order is greater than three. Not only is this an unnecessary restriction for students with access to calculators or computers, but also it is certain to give students very little understanding of how to solve such systems in general or of the pitfalls that may occur when trying to solve such systems.

It is quite appropriate to begin discussing linear systems by giving geometric illustrations of what it means to solve 2×2 and 3×3 systems and then to solve these systems algebraically in an ad hoc manner in order to lead students toward the algorithmic generalization. An important point here though is that for 2×2 systems, the elimination should not be presented by multiplying both equations by suitable constants and then subtracting one from the other. Instead, the correct approach, not because it is better for 2×2 systems but because it is much more effective for larger systems, is to subtract the appropriate multiple of one equation from another. For example, to solve

$$3x - 4y = 1$$
$$2x + 3y = 12,$$

it is easiest to multiply the first equation by 2 and the second by 3 and then subtract to get

$$-17y = -34,$$

but it is more generally applicable to subtract ⅔ of the first equation from the second to obtain

$$17y/3 = 34/3.$$

Of course, multiplying each of the two equations by a constant and then subtracting is a technique that is practicable only when all the coefficients are integers or simple rational numbers. Another advantage of the second approach is that there is no need to give students the misimpression that equations always have simple coefficients.

The algorithm in figure 24.3 is the bare-bones form of Gaussian elimination. To understand this figure, you must express the system using subscript notation in the standard form

$$a_{11}x_1 + a_{12}x_2 + \ldots + a_{1n}x_n = b_1$$

$$\vdots$$

$$\vdots \qquad\qquad\qquad\qquad\qquad\qquad (7)$$

$$\vdots$$

$$a_{n1}x_1 + a_{n2}x_2 + \ldots + a_{nn}x_n = b_n.$$

For students who have applied the elimination method to 2×2 and 3×3 systems and who are comfortable with the notation in (7), understanding the algorithm in figure 24.3 should not be difficult. One advantage of the approach of figure 24.3 is that it facilitates discussion of various pitfalls. For example, what happens when one of the coefficients a_{ii} is zero?

```
Input A, b                      [Matrix of coefficients and right-hand-side vector]
Algorithm GaussElim
        for i = 1 to n−1                        [Eliminate variables in column i]
          for k = i+1 to n
            m_k ← a_ki/a_ii          [Multiple of row i to subtract from row k]
            a_ki ← 0                              [Element eliminated]
            for j = i+1 to n                      [New values in row k]
              a_kj ← a_kj − m_k a_ij
            endfor
            b_k ← b_k − m_k b_i
          endfor
        endfor
Output Triangular matrix of coefficients and corresponding right-hand sides
```

Fig. 24.3. Algorithm for Gaussian elimination

APPLYING THE RECURSIVE AND INDUCTIVE PARADIGMS

The recursive paradigm (RP), which we introduced in the first section, often strikes most people as quite foreign, even though it is really only a specialized version of "reduce to the previous case." Horner's method provides a nice illustration.

Suppose a student wonders how you would be able to discover the form (4) if you didn't know it already. *Suppose you already know* how to evaluate polynomials of degree $n - 1$. How can you use this knowledge to evaluate a polynomial of degree n? If you were using the brute-force approach, you would just add $a_n x^n$ to the polynomial of degree $n - 1$. This is a recursive approach, but using the polynomial of degree $n - 1$ as the previous case is not nearly as clever as the choice in Horner's method. Using Horner's method, rewrite the polynomial as

$$x(a_n x^{n-1} + \ldots + a_1) + a_0. \tag{8}$$

The polynomial in parentheses is a polynomial of degree $n - 1$, which we have assumed you know how to evaluate. To evaluate the polynomial of degree n, multiply by x and add a_0. Now continue by realizing that if you can evaluate a polynomial of degree n by reducing to an evaluation of a polynomial of degree $n - 1$, then you can evaluate the polynomial of degree $n - 1$ by reducing to an evaluation of a polynomial of degree $n - 2$. This permits (8) to be rewritten as

$$x(x(a_n x^{n-2} + \ldots + a_2) + a_1) + a_0.$$

Continuing in this way, you arrive at (4).

This idea—jumping right into a typical case, supposing that you know how to treat a previous case, and working your way down and back—is the essence of *recursion*. Starting from the original polynomial and getting to (4) is the *down* stage. Then evaluating (4) is the *back* stage; you start with the initial case $(a_n x + a_{n-1})$ and work up to the whole polynomial.

The development of the algorithm for Gaussian elimination furnishes another illustration of the RP, which we will sketch briefly. Suppose you already know how to solve $(n-1) \times (n-1)$ systems and want to solve an $n \times n$ system. Subtracting the appropriate multiple of the first equation of (7) from each of the other equations to eliminate x_1 from each and, for the moment, ignoring the first equation, you obtain an $(n-1) \times (n-1)$ system, which, it is assumed, you can solve for $x_2, x_3, \ldots, x_n$. Substituting back into the first equation, you get x_1. Apply this idea to the $(n-1) \times (n-1)$ system to solve it in terms of an $(n-2) \times (n-2)$ system. Working down in this way, you arrive finally at one equation in one unknown. Try it!

We have discussed recursion as a discovery technique, but it is also a computational technique. Indeed, most programming languages allow recursive procedures that can invoke themselves. This makes it possible for recursive relationships to be translated directly into computational procedures. However, since recursive procedures take some getting used to and because it often increases efficiency to implement recursion using iterative constructs, we have presented Horner's method and Gaussian elimination in iterative form. But there are other problems such as those found in chapters 18–20 of this yearbook where recursion is the only straightforward computational approach.

The inductive paradigm (IP), also introduced in the first section, should seem familiar; it is essentially the discovery approach to mathematics, so often touted but so seldom used. Let us apply the first two steps (compute, conjecture) to the polynomial multiplication problem.

Thus, consider

$$\left.\begin{array}{l}
(a_1 x + a_0)(b_1 x + b_0) = a_1 b_1 x^2 + (a_1 b_0 + a_0 b_1)x + a_0 b_0 \\
(a_1 x + a_0)(b_2 x^2 + b_1 x + b_0) \\
\quad = a_1 b_2 x^3 + (a_1 b_1 + a_0 b_2)x^2 + (a_1 b_0 + a_0 b_1)x + a_0 b_0 \\
(a_2 x^2 + a_1 x + a_0)(b_2 x^2 + b_1 x + b_0) \\
\quad = a_2 b_2 x^4 + (a_1 b_2 + a_2 b_1)x^3 \\
\quad + (a_2 b_0 + a_1 b_1 + a_0 b_2)x^2 + (a_1 b_0 + a_0 b_1)\, x + a_0 b_0
\end{array}\right\} \quad (9)$$

These examples should be enough to suggest to the student that the coefficient of x^i in the product polynomial is the sum of products of coefficients whose subscripts add to i.

The IP is also useful in teaching mathematical induction where the usual approach is to give the result ("mathematics as magic") and then prove it by induction. For example, instead of stating and then proving that the sum of the first n positive integers is $n(n + 1)/2$, it is preferable to begin by displaying a table of sums for, say, n going from 1 to 10 and then to conjecture the result by analyzing the pattern in the table.

ALGORITHMS AND PROOFS

Developing an understanding of what a proof is and developing facility with constructing proofs must remain an important goal of high school mathematics. The quintessential proof technique in discrete mathematics is mathematical induction, already discussed in chapter 21 of this volume. Algorithms provide an excellent medium for discussing proofs by mathematical induction.

For example, consider the algorithm for Horner's method in figure 24.1. Is it really correct? What is it that we wish to have computed each time the **for . . . endfor** loop is executed? After the first time through the loop

$$p = a_n x + a_{n-1}, \tag{10}$$

and after the second time

$$p = x(a_n x + a_{n-1}) + a_{n-2}$$
$$= a_n x^2 + a_{n-1} x + a_{n-2}.$$

In general (inductive paradigm!), after k passes through the loop

$$p = \sum_{i=0}^{k} a_{n-i} x^{k-i}. \tag{11}$$

The quantity in (11) is called a *loop invariant* because—

- it gives the value of p on the first entry to the loop in figure 24.1 (when k is implicitly equal to 0; with $k=0$, (11) gives $p = a_n$);
- it also gives the value of p on each subsequent exit from the loop (e.g., after the first pass p is given by (10), which is the same as (11) with $k=1$).

More generally, if (11) is true after the kth pass through the loop, it is also true after the $(k + 1)$st pass. This is so because on the $(k + 1)$st pass p is multiplied by x and a_{n-k-1} is added to get the new p. Performing these same operations on (11), we get precisely the right-hand side of (11) with k replaced by $k+1$. Thus, after n passes through the loop we have (11) with $k = n$, which is just the polynomial itself evaluated at x.

What we have just done, in effect, is to give a proof by induction that

Horner's algorithm does what it purports to do. The proposition $P(k)$ is that p, as given by (11), is the value after k passes through the loop. In the previous paragraph we showed that the basis case ($k=0$) is true and that $P(k) \Rightarrow P(k+1)$. Since $P(n)$ is the statement that, after the final exit from the loop (which is after the nth pass through it), the value of p, as given by (11) with $k = n$, is that of the polynomial evaluated at x, this *finite* induction verifies Horner's algorithm. This technique can be used to prove many algorithms correct.

THE ANALYSIS OF ALGORITHMS

One of the great charms of an algorithmic approach to mathematics is that it enables us to analyze effectively how efficient a method is and to compare one algorithm for a task with another for the same task.

Efficiency considerations apply to even so basic a subject as the distributive law—$ax + bx = (a + b)x$. How many arithmetic operations are required for each side of the equation? How many button pushes on your calculator? Factoring certainly is the best method in answer to the first question. For the second, the answer depends on whether your calculator requires parenthesis buttons and on the button pushes to input x.

Polynomial evaluation furnishes a more sophisticated example. How many additions and multiplications are required by Horner's method? From figure 24.1 the answer is seen to be n of each. How does this compare with the brute-force approach of computing the powers of x from 2 to n and then multiplying by the coefficients and adding? Computing the powers requires $n-1$ multiplications (assuming that you save each power for later use), multiplying by the coefficients requires n more, and there are n additions. So the number of additions is the same as with Horner's method, but the number of multiplications is $2n - 1$, which justifies the preference for Horner's method over brute force. Analyses like these can be made for quadratic equation solving (Ralston 1985) and for the other problems discussed in this chapter. Asking students to work on these will result in some good class discussions.

ALGORITHMS AND THE SECONDARY SCHOOL CURRICULUM

Our examples so far have been of topics currently in the standard curriculum whose presentation can be enhanced by using an algorithmic approach. But there are large areas of discrete mathematics that rarely if ever appear in the secondary school curriculum that should be candidates for inclusion and where an algorithmic approach is valuable if not mandatory.

One example is graph theory. Indeed, some classical graph theorems are already familiar as enrichment topics, for example, Euler's theorem that a connected graph is traversable without repeating edges if every vertex is incident to an even number of edges. But, this is an *existence* statement. How do you *construct* such a traversal? In almost any real application (and there are many), the graph is too large even to draw much less to eyeball a traversal. But there are algorithms, and students can be led to discover them.

Here is another graph theory example. Suppose a developing country wants to link all its major cities by a new telephone network (fig. 24.4). It does not need a direct link between each pair of cities but only a *path* between any two cities. If the cost to build a link between any two cities is known, then a *minimum spanning tree* provides the lowest cost network. An algorithm to construct this tree is easy to discover and is a naive *greedy* approach. Choose first the edge with the smallest cost; then at each step choose the edge with the smallest cost that joins a vertex already in the tree to one that is not in the tree; break ties arbitrarily. (Justifying that this works is quite a bit harder, however.)

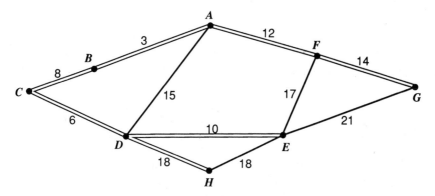

The numbers on the edges represent costs. The double line indicates the minimum spanning tree (i.e., the acyclic graph that contains every vertex and whose sum of edge weights is the minimum). The order in which the edges are added is (A,B), (B,C), (C,D), (D,E), (A,F), (F,G), (D,H).

Fig. 24.4. A weighted graph and its minimum spanning tree

Counting permutations and combinations has long been a standard secondary school topic. But it is nonstandard to discuss how to generate one or more permutations or combinations at random. However, for many applications, particularly statistical ones, this is the real need. It is easily done using the RP. Suppose you already knew a random permutation of n items. Randomly pick a *position* for the $(n + 1)$st item and insert it into the n-permutation at the generated position. Various aspects of this problem offer good exercises for student discovery. For example, how do you get a permutation of k things out of n efficiently? How do you generate permutations on a computer instead of by hand, since inserting something into a list on a computer is trickier than just putting it into a space left on a piece of paper?

SOME FINAL THOUGHTS

The algorithmic notation used in this article is intended to have three properties:

1. It is *informal* enough for it to be readable after only a very brief introduction to the syntax and meaning of the various constructs. Both mathematical notation and English should be usable wherever convenient.

2. It is *formal* enough for the conversion of an algorithm to a computer program in any common programming language to be fairly straightforward.

3. It is precise enough for students to understand exactly how algorithms work and for their properties to be proved without excessive handwaving (i.e., vagueness).

We call this notation *algorithmic language*. Sometimes computer scientists call it "pseudocode," but from the standpoint of communication between people rather than computers, there is nothing "pseudo" about it. Algorithmic notation is recommended in the *Curriculum and Evaluation Standards for School Mathematics* (NCTM 1989); the specific algorithmic language in this chapter is used extensively in Maurer and Ralston (1991) in many examples of algorithms for tasks in discrete mathematics.

Can you teach discrete mathematics without using algorithms? Despite our title we suppose you *can*. But we are convinced that you cannot teach discrete mathematics *well* without using an algorithmic approach. A focus on algorithms allows you to achieve a level of mathematical breadth, depth, and relevance that is almost impossible without using them.

REFERENCES

Maurer, Stephen B. "Two Meanings of Algorithmic Mathematics." *Mathematics Teacher* 77 (September 1984): 430–35.

Maurer, Stephen B., and Anthony Ralston. *Discrete Algorithmic Mathematics*. Reading, Mass.: Addison-Wesley Publishing Co., 1991.

National Council of Teachers of Mathematics. *Curriculum and Evaluation Standards for School Mathematics*. Reston, Va.: The Council, 1989.

Ralston, Anthony. "The Really New College Mathematics and Its Impact on the High School Curriculum." In *The Secondary School Mathematics Curriculum*, 1985 Yearbook of the National Council of Teachers of Mathematics, edited by Christian R. Hirsch. Reston, Va.: The Council, 1985.

25

Anagrams: An Application of Discrete Mathematics

Angela B. Shiflet

THE advent of computers not only has prompted the study of discrete mathematics but has also contributed many interesting examples to the subject. One such application of discrete mathematics is *anagrams*, which are different words that use the same letters. For instance, the words *below*, *elbow*, and *bowel* are anagrams, as are *large* and *regal*.

Classes on many different grade levels, from elementary to college, can explore various questions involving anagrams and mathematics. In trying to discover a good method for using the speed of a computer to find all the anagrams in English, students use discrete mathematical concepts from the fundamental theorem of arithmetic and combinatorics to recursion and binary trees. Moreover, the development of this unusual application of discrete mathematics involves the calculator, estimation, logarithms, units analysis, and algorithms. Its study can even strengthen vocabulary and spelling!

An enjoyable exercise for an elementary school class is to see who can find the most anagrams or the most words that are anagrams of each other. Or give the class a word and let the students use the dictionary to find all the anagrams of the word along with their definitions.

Dictionaries can also be stored on the computer. Perhaps some of your students have used a word processor with a built-in dictionary and a spelling checker. After the class has worked on the game for a while, pose the question, "If we could work as fast as a computer, what method, or *algorithm*, could we use to find all possible anagrams in the dictionary?"

Undoubtedly, one student will suggest taking the words one at a time and looking up all possible rearrangements of the letters. Such a suggestion can lead to a discussion of *permutations*, or ordered arrangements. A calculator can then be used to find the number of permutations of the twelve-letter word *ambidextrous*. Ask students to estimate, "If our computer can check one arrangement of letters every tenth of a second, how long will it take the computer to find all the anagrams of this word?" The students' estimates will vary widely, but a computation on the calculator with units analysis reveals the startling result that it will take our computer about one-and-a-half years to check all the permutations of this one word:

$$12! \text{ permutations} = 479\ 001\ 600 \text{ permutations}$$

$$479\ 001\ 600 \text{ perms} \times \frac{0.1 \text{ s}}{1 \text{ perm}} \times \frac{1 \text{ min}}{60 \text{ s}} \times \frac{1 \text{ h}}{60 \text{ min}}$$

$$\times \frac{1 \text{ day}}{24 \text{ h}} \times \frac{1 \text{ year}}{365 \text{ days}} = 1.5 \text{ years}$$

Another idea for finding all possible anagrams is to take each word and go through the dictionary word by word looking for others that contain the same letters. Since we do not need to search for each permutation but can go through the list sequentially, we might expect this algorithm to be workable with a fast computer. Unfortunately, this brute-force method also takes a prohibitive amount of time. Suppose our dictionary contains $n = 150\ 000$ words. Using combinatorics again, we see that our method will take on the order of $n^2 = 225 \times 10^8$ comparisons. We can ask, "If our extremely fast computer can make 100 000 comparisons of words per second, how long will it take to locate all the anagrams?" After fielding estimates from the class, we grab our calculators and discover that the work will take over two-and-a-half days.

$$225 \times 10^8 \text{ comparisons} \times \frac{1 \text{ s}}{10^5 \text{ comparisons}}$$

$$\times \frac{1 \text{ min}}{60 \text{ s}} \times \frac{1 \text{ h}}{60 \text{ min}} \times \frac{1 \text{ day}}{24 \text{ h}} = 2.6 \text{ days}$$

CRALLE'S ALGORITHM

A clever method by Robert K. Cralle (1982) of Lawrence Livermore National Laboratory uses the fundamental theorem of arithmetic to develop a much faster algorithm. Start by associating a prime number with each letter, as follows:

a	b	c	d	e	f	g	h	i
2	3	5	7	11	13	17	19	23

j	k	l	m	n	o	p	q	r
29	31	37	41	43	47	53	59	61

s	t	u	v	w	x	y	z
67	71	73	79	83	89	97	101

After associating each letter with a different prime in the algorithm to find all anagrams, the computer calculates the product of the corresponding primes for each word. For example,

below → 3 × 11 × 37 × 47 × 83 = 4 763 121,
elbow → 11 × 37 × 3 × 47 × 83 = 4 763 121.

The fundamental theorem of arithmetic states that each positive integer factors in exactly one way (other than possibly different orders) into a product of primes. The only factors of 4 763 121 are the primes 3, 11, 37, 47, and 83; and only the words with the associated number 4 763 121—such as *below, elbow,* and *bowel*—are anagrams of each other.

After pairing each word with its product, the computer sorts the list into ascending order by the number. Since anagrams share the same number, all the anagrams of a word will be clustered with that word in the sorted list.

If our topic is permutations or the fundamental theorem of arithmetic, we might stop the example here; but if we are teaching algebra 2, we can use this work to illustrate another concept—equivalence classes. We can define two words as being related to each other if their corresponding numbers are equal, that is, if the words are anagrams. The class can verify that this relation is in fact an equivalence relation. Moreover, after sorting, all the words in an equivalence class appear together.

We can also use this example in a discussion of logarithms. Is this algorithm using the fundamental theorem of arithmetic really faster than the two brute-force methods discussed earlier? The computer must calculate a product for each of the n words in the dictionary, and, perhaps, $n = 150\ 000$. This process will take time on the order of n; that is, the time to complete the multiplications associated with n words is proportional to n. Even if the computer can calculate only a thousand products in a second, the computations will take only 150 seconds, or 2.5 minutes. Arranging the list in ascending order, however, is significantly more time consuming. It has been shown that the fastest sorting techniques that use comparisons are on the order of $n \log_2 n$; that is, the number of comparisons and, consequently, the time for such a sorting of n items is proportional to $n \log_2 n$. The actual time, however, varies from one computer to another. Using calculators, the class can evaluate $\log_2 150\ 000$ as

$$\log_2 150\ 000 \ = \ \frac{\log_{10} 150\ 000}{\log_{10} 2} \ = \ 17.1946.$$

Thus, the sorting process uses on the order of

$$n \log_2 n \ = \ 150\ 000 \times 17.1946 \approx 2\ 579\ 190$$

comparisons. With the second brute-force method above, the number of comparisons is on the order of $n^2 = 22\ 500\ 000\ 000$ comparisons. Compute

$$\frac{22\ 500\ 000\ 000}{2\ 579\ 190} = 8724$$

to see that our new technique is almost nine thousand times faster than the brute-force one. We do in fact have a better algorithm.

The calculations show that some algorithms are too slow for the fastest supercomputers. A clever application of discrete mathematics may mean the difference between a project that is possible to do and one that is not.

A SORTING ALGORITHM

The curious student should wonder how to do the fast sorting needed in the last algorithm. There are several such methods, one of which leads to a discussion of binary trees. Before we begin the sorting, we must cover some of the terminology of graph theory. A *tree* is a set of *vertices* and a set of *edges* joining pairs of vertices, such that the drawing is in one piece and such that we cannot start at a point and travel in a circular fashion through the tree back to the same point. As figure 25.1 shows, a tree resembles an upside-down version of its live counterpart. In this figure the top vertex, *a*, which is called the *root* of the tree, has two *children*, *b* and *c*. The tree is a *binary tree* because every vertex has at most two children. Vertex *b* is the

Fig. 25.1. Example of a binary tree

root of the *left subtree* of *a*, just as *c* is the root of the *right subtree*. We say that *s* is a *subtree* of a tree *t* if *s* is itself a tree and every vertex and edge of *s* are in *t*. (See fig. 25.2.)

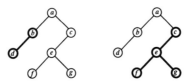

Fig. 25.2. Left and right subtrees (**in bold face**) of the binary tree in figure 25.1

There are several methods that use trees for sorting a list. In one technique, we place each item in a vertex of a special tree called a *binary search tree*, and then travel through the tree in a way that yields the list in ascending order. If during the building process the tree remains "bushy" with each subtree having approximately the same number of nodes on the left as on the right, then the speed of this technique is on the order of $n \log_2 n$.

To illustrate the process, let us place the numbers 6, 3, 8, 2, and 4 in a binary search tree. The procedure to place a value into a binary search tree is *recursive*; that is, it calls itself. When inserting a number into a tree, we may need to use the same process to insert the number into a subtree. After the tree is complete, we can read the numbers in ascending order.

The first number, 6, is placed in the root as shown in step 1 of figure 25.3. If the next number is less than the root number, we install it as the left child. If the number is greater, we place it as the right child. Thus, as step 2 shows, the number 3 becomes the left child, and the next number, 8, goes to the right (step 3). The number 2 is less than the value at the root, 6, so we send 2 to the left; but another value is there. Consequently, in a recursive manner, we repeat the process with the left subtree. Since 2 is less than 3 (the value in the root of the subtree), in step 4 we insert 2 as the left child of 3. Similarly, $4 < 6$; so we continue the comparison with the left subtree. Since $4 > 3$, we install 4 as the right child of 3 (step 5).

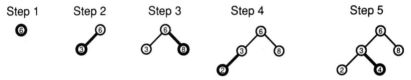

Fig. 25.3. Insertion of the numbers 6, 3, 8, 2, and 4 into a binary search tree

The recursive algorithm to insert value e into a binary search tree t can be described as follows:

InsertBST(t, e):
If the tree t is empty then
         place the value e in the root
else if e is less than the value in the root then
         InsertBST(the left subtree of t, e)
else
         InsertBST(the right subtree of t, e)

We see that this algorithm is recursive, calling itself. If during the insertion of e we encounter a vertex, we go to either the left or the right and repeat the whole process, this time inserting e into the proper subtree. Although the procedure calls itself, it performs the insertion on a smaller tree. More-over, we have a way to stop—when we encounter an empty subtree, we place the value e into a vertex at that location.

A RETRIEVAL ALGORITHM

Once the binary search tree has been built, we are ready to retrieve the list in ascending order. To accomplish this objective, we *traverse* or travel through the tree in a certain way, called an *inorder traversal*. As we traverse the tree, periodically we will *visit* a vertex or perform a certain procedure using the information stored in that vertex. For the purpose of this example the visit procedure will print the value in the vertex. Like *InsertBST,* the algorithm for an inorder traversal of the tree is recursive.

InorderTraversal(t):
    If the root of the tree is not empty then
        InorderTraversal(the left subtree of *t*)
        Visit the root
        InorderTraversal(the right subtree of *t*)

For the inorder traversal of the binary tree in figure 25.4 (from step 5 of fig. 25.3), we cover the vertices in the sorted order: 2, 3, 4, 6, 8. In the following detailed description of the traversal of this tree, we use boldface to emphasize a visit and indentation to indicate statements that are part of the traversal of one subtree:

InorderTraversal of Binary Tree in figure 25.4:
Traverse the left subtree of 6
    Traverse the left subtree of 3
        Traverse the left subtree of 2—empty, do nothing
        Visit the root 2
        Traverse the right subtree of 2—empty, do nothing
    Visit the root 3
    Traverse the right subtree of 3
        Traverse the left subtree of 4—empty, do nothing
        Visit the root 4
        Traverse the right subtree of 4—empty, do nothing
Visit the root 6
Traverse the right subtree of 6
    Traverse the left subtree of 8—empty, do nothing
    Visit the root 8
    Traverse the right subtree of 8—empty, do nothing

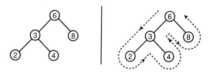

Fig. 25.4. A binary tree drawn without and with a dotted curve to aid traversal

We still have not convinced the curious student that this method of sorting for certain data sets is on the order of *n* $\log_2 n$. Suppose we wish to insert the value 5 into the binary search tree of figure 25.4. Using the process defined above, we need to compare 5 to only three values—6, 3, and 4—before finding the proper location. If we wish to insert 7, only two comparisons are needed, one with the root, 6, and another with 8. For this tree of five vertices that is fairly well balanced, the maximum number of comparisons is three. This is the *ceiling* of $\log_2 5$, that is, the smallest integer that is greater than or equal to $\log_2 5$.

For another example, consider the balanced tree in figure 25.5. To insert any vertex into the tree we must make a maximum of four comparisons, the ceiling of $\log_2$ 15. Of course, if the tree is not so well balanced, the number of comparisons can be greater than $\log_2 n$. To place a value into a "bushy" binary search tree of n vertices, however, the number of comparisons is about $\log_2 n$. Since we must repeat the insertion process for each of the n values, the total building process for a tree that remains fairly balanced is on the order of $n \log_2 n$. Once the tree is built, traversing it to write the values in order is very fast, on the order of n. Thus, the total building and traversal process takes about $n \log_2 n + n$ comparisons. For the example of a dictionary with $n = 150\ 000$ words, we saw that $n \log_2 n = 2\ 579\ 190$, a value seventeen times that of n. Since for a lot of data $n \log_2 n$ dominates the value of n, we say the sorting process is on the order of $n \log_2 n$.

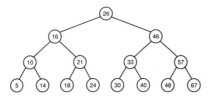

Fig. 25.5. A balanced tree of fifteen vertices

SUMMARY

Finding all the anagrams in the English language can be interesting and instructive from many different perspectives and levels of difficulty. Discovering the number of ordered arrangements of the letters of a word illustrates permutations and factorials. Investigating the time to find anagrams reinforces estimation skills and calculator use and employs logarithms and units analysis. Applying the first step of the Cralle algorithm for finding all the anagrams uses primes and the fundamental theorem of arithmetic. Completing the method by sorting allows us to study binary search trees and employ recursion in their development and traversal. The sorted list shows us an example of equivalence classes. Thus, the application of discrete mathematics to anagrams can illustrate topics in a wide variety of mathematics courses. Because of its many applications, discrete mathematics can and should occur throughout the mathematics curriculum.

BIBLIOGRAPHY

Cralle, Robert K. "Brevity Is the Soul of Programming." *Tentacle* 2 (December 1982): 25–27.

Shiflet, Angela B. *Discrete Mathematics for Computer Science.* Saint Paul, Minn.: West Publishing Co., 1987.

_____. *Elementary Data Structures with Pascal.* Saint Paul, Minn.: West Publishing Co., 1990.

26

An Introduction to Game Theory

Kevin G. Bartkovich

A LOCAL fast-food restaurant is deciding on a marketing strategy. The United States is making decisions on whether to disarm or deploy additional missiles. Seniors at a high school are deciding whether to allow underclassmen to attend the prom. These situations can be analyzed mathematically using techniques from a field of mathematics called *game theory.*

The word *game* in game theory refers to a situation in which two or more participants, called *players,* have conflicting interests, and each player has partial but not total control over the outcome of the conflict. We seek to construct a mathematical model for a game by quantifying the possible outcomes of a conflict and determining the best strategies for the players.

Game theory is a worthwhile topic for secondary school students for several reasons. Situations that can be analyzed and illuminated with game theory are commonplace; so the mathematics is relevant to a student's world. At the beginning level, no advanced mathematical concepts are needed to understand game theory other than finding the intersection of lines in the plane. With a brief instructional time, students can investigate sophisticated conflicts without needing much formal mathematics. The problems encountered in game theory are open-ended, leading to debate and discussion among students. Modern game theory is a recent development; most of the relevant mathematics has been developed in the last fifty years. The study of game theory gives students a sense that mathematics is dynamic, current, and applicable to everyday life.

ZERO-SUM GAMES

A Food Fight

In the town of Market Square, the only restaurants are Mike's Steak House and Sally's Seafood. The town spends, on the average, $10 000 a month dining out. The two restaurants split this money, so customers at Mike's are lost sales for Sally's and vice versa. Prior to the beginning of each month, the owners decide on marketing strategies for that month. They each have four choices: do nothing different from the standard menu, introduce a new item on the menu, offer a special price for a combination meal (for example, a meat and two vegetables), or offer a free dessert with any meal. Each combination of strategies by the two restaurants will yield a payoff in sales for each. There are sixteen possible combinations of strategies and therefore sixteen possible payoffs. A marketing firm has given the restaurants the list of projected payoffs in figure 26.1. Each payoff represents, from the perspective of Mike's Steak House, the projected difference in the revenues between the restaurants. For example, if both do nothing different from the standard menu, the money for that month is split with $6000 going to Mike's and $4000 going to Sally's; hence the entry in the payoff matrix is 2 (the difference between the revenues, in thousands). A payoff element of -6 implies that Mike's earns $2000 while Sally's earns $8000. A payoff element of 0 denotes an even split of $5000 for each restaurant. The other payoff elements are interpreted in an analogous manner.

| | | Sally's Seafood | | | |
		Nothing	New item	Special	Dessert
	Nothing	2	-3	-6	-4
Mike's	New item	-3	4	-2	0
Steak House	Special	5	2	6	1
	Dessert	7	-2	-2	-1

Fig. 26.1. Payoff matrix for two restaurants

This game is an example of a *two-person game*, in which one player is pitted against a second player. The *row player*, which is Mike's Steak House in this situation, has possible strategies represented by the rows of the *payoff matrix*. The *column player*, Sally's Seafood, has possible strategies represented by the columns of the payoff matrix. The assumption that one player's gain is the other player's loss places this game in a general class called *zero-sum games*. As is done here, a game-theory payoff matrix is constructed from the perspective of the row player. The payoffs from the column player's perspective are the negatives of each entry in the matrix.

Game theory seeks to answer the following question: Given that Mike's has to pick a marketing strategy each month, what strategy should be chosen in any given month? First, notice that Mike's "do nothing" strategy (row 1) can always be bettered by choosing the "special" strategy (row 3). No matter what strategy Sally's chooses, Mike's payoff in row 3 is greater than the payoff in row 1. This property is called *row domination*, and row 3 is said to *dominate* row 1. In fact, row 1 can be removed from the payoff matrix, since Mike's will never choose it over row 3. There is no column dominance, so Sally's cannot eliminate any strategy as easily.

In the process of choosing a marketing strategy, the owner of Mike's might take an optimistic view: "If I decide to play row 4 (free dessert) and Sally's chooses to play column 1 (do nothing), I will get my maximum sales this month." Although this statement is perhaps overly optimistic, suppose that for several months in a row this exact scenario is played out. Mike's plays row 4 and Sally's chooses column 1. Eventually the owner of Sally's will catch on to Mike's strategy. What will happen next? Sally's will switch strategies from column 1 to column 3 (special), giving Sally's $2000 more in sales than Mike's this month. After seeing this occur a few times, Mike's owner will not stand by and lose money. By switching strategy from row 4 to row 3 (special), Mike's can obtain $6000 more in sales than Sally's in a month. If Mike's continues to offer a special in succeeding months, what is Sally's best response? Sally's would do best by switching strategies from column 3 to column 4 (free dessert). Sally's would still earn less than Mike's ($1000 a month), but any other strategy would earn even less money as long as Mike's continues to play row 3.

The game will not move from this payoff in row 3 and column 4. Mike's can do no better in column 4 by switching from the row 3 strategy. Sally's can do no better in row 3 by switching column strategies. The game has reached an equilibrium point.

What if the game begins with row and column strategies different from the row 4, column 1 combination we started with above? Repeated play such as in the scenario above reveals that no matter where the game starts in the payoff matrix, it will end up at the equilibrium point in row 3 and column 4, which has a payoff of 1. This payoff is called the *value* of the game. An equilibrium point such as this in a zero-sum game is called a *saddle point*. The saddle-point entry in a matrix is the smallest in its row and the largest in its column, much as the seat of a saddle is simultaneously a minimum and maximum depending on the perspective of the observer.

The existence of a saddle point can be determined by examining the *play-safe* strategies of each player. A play-safe strategy assumes that the worst-case scenario will occur, and so a player looks for the best of the worst-case outcomes. The play-safe strategy for the row player is determined by examining the worst-case outcome for each row, which is the minimum entry in each row. The minimum row entries are -6 for row 1, -3 for row 2, 1 for row 3, and -2 for row 4. The row player's play-safe strategy seeks to *maximize* the worst-case outcome, which is accomplished by choosing row 3 for a worst-case payoff of 1. The worst-case outcomes for the column player, by contrast, are the maximum entries in each column (recall that the payoffs for the column player are the negative of the entries in the payoff matrix). The maximum column entries are 7 for column 1, 4 for column 2, 6 for column 3, and 1 for column 4. The column player's play-safe strategy is to choose the smallest of the worst-case outcomes, which is the worst-case payoff of 1 for column 4. The two players' play-safe strategies yield equal payoffs for the game, indicating that the maximum of the minimum row payoffs equals the minimum of the maximum column payoffs. The location of the entry in the payoff matrix where this occurs is the saddle point. The matrix in figure 26.2 illustrates the preceding analysis.

| | | Sally's Seafood | | | | |
		Nothing	New item	Special	Dessert	Row minimum
	Nothing	2	−3	−6	−4	−6
Mike's	New item	−3	4	−2	0	−3
Steak House	Special	5	2	6	1	1
	Dessert	7	−2	−2	−1	−2
Column maximum		7	4	6	1	

Fig. 26.2. Existence of a saddle point in a payoff matrix in thousands of dollars

The analysis of the food fight game depends on three assumptions that are stated below. These assumptions form the basis for game theory.

Assumptions of game theory

1. Each player has identical information about the payoff matrix, allowing predictions about what the opposing player will do.
2. Each player plays the game in a rational and competent manner.
3. Each player seeks to maximize his or her payoff in the game.

The analysis of the game between Mike's and Sally's assumes that the game is played repeatedly month after month. If the game was played only once, the existence of a saddle point would not necessarily determine the outcome of the game. In any single play, a player may gamble and not choose the saddle-point strategy. For example, Mike's owner might choose row 4 in a one-time play of the game because of the chance of obtaining a payoff of 7. Likewise, Sally's owner might choose column 3 in the hope of obtaining a payoff of 6. With repeated play, if one player deviates from the saddle-point strategy, then over the long run that player will obtain a worse payoff than at the saddle point. With repeated play, each player is motivated to use the saddle-point strategy, even from the first play of the game.

A player choosing the same row or column throughout repeated plays of a game is said to be using a *pure strategy*. In the next game, a saddle point does not exist, and the players do not use pure strategies.

Pitcher versus Batter

Dizzy Dan is on the pitcher's mound facing the feared slugger Hammerin' Hank. Dizzy has two strong pitches in his arsenal, a fast ball and a curve ball. Knowing that Dizzy will throw one of these pitches, Hank seeks to maximize his chances of getting a hit by guessing what Dizzy will throw. Of course, if Hank guesses correctly, he will do much better than if he guesses incorrectly. The team statistician has compiled Hank's batting average against curve balls and fast balls relative to when he guesses one or the other. These numbers are given in the matrix shown in figure 26.3.

| | | Dizzy's throws | |
		Fast ball	Curve ball
Hank's	Fast ball	.400	.250
guesses	Curve ball	.150	.300

Fig. 26.3. Payoff matrix for pitcher versus batter

This game is zero-sum because we can assume that the payoff in hits for Hank is to the detriment of Dizzy in relatively the same amount. Examining the play-safe strategy for the row player, we find that the minimum entries in the two rows are .250 and .150; therefore, Hank's play-safe strategy is to guess fast ball (row 1) with a worst-case value of .250. The maximum entries in the columns are .400 and .300, so Dizzy's play-safe strategy is to throw the curve ball (column 2) with a worst-case value of .300. Since the play-safe values are different for Hank and for Dizzy, this game does not have a saddle point, and, therefore, it does not settle down to an equilibrium point. For example, if Hank continually guesses fast ball and Dizzy throws curve balls, then it would benefit Hank to start guessing curve ball (making his batting average go from .250 to .300). But then Dizzy would switch to a fast-ball strategy, thus lowering Hank's chance of getting a hit to .150, which in turn would motivate Hank to guess fast ball, so that his batting average would rise to .400. This strategy-switching could continue indefinitely, never stabilizing at a particular entry in the matrix.

The cycle of strategies in this game demonstrates that if player 1 can predict what strategy player 2 will use, then that knowledge can be used to maximize the gain of player 1. The notion that players will look for predictable patterns in each other's play will lead each player to pick a strategy at random for each play of the game. A question for consideration is what kind of randomization device to use. Flipping a coin may work, but it leads to an approximate 1-to-1 ratio for the number of times each of two strategies is chosen. Perhaps a player would do better by choosing between two strategies so that strategy A is chosen half as often as strategy B, although still at random. (A 2-to-1 ratio could be achieved by throwing a fair die. If 5 or 6 appears, then choose strategy A; otherwise, choose strategy B.)

The process of randomly choosing between alternatives is called playing a *mixed strategy*. What are the best mixed strategies for Hank and Dizzy? In other words, how should they randomize their choices so that they achieve maximum payoffs over repeated plays of the game? We first examine the game from Dizzy's perspective. Let us assign a probability p to his throwing a fast ball (column 1). This implies that he throws a fast ball a proportion p of the time over the long run. In addition, this means that he throws a curve ball (column 2) with probability $1 - p$.

Suppose that Hank plays a pure strategy of guessing fast ball (row 1) while Dizzy is playing this mixed strategy. The value of the game is the sum of the outcomes in row 1 multiplied by the probabilities that each occur. Thus, if Hank plays row 1, then the value of the game is

$$.400p + .250(1 - p).$$

If $p = 1$, then the value of the game is .400, and if $p = 0$, then the value of the game is .250, as we would expect by referring to the payoff matrix.

Suppose Hank chooses to play only row 2, always guessing curve ball. In this case the value of the game is

$$.150p + .300(1 - p).$$

If $p = 1$, then the value of the game is .150, and if $p = 0$, then the value of the game is .300.

Each of the two expressions for the value of the game above are linear functions of p. With p varying between 0 and 1, we can sketch both of these lines, as shown in the graph in figure 26.4.

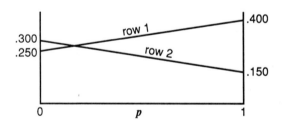

Fig. 26.4. Value of the game for each of Hank's pure strategies

Since the entries in the payoff matrix are from Hank's perspective, he would like to make the value of the game as large as possible. If Hank plays a pure strategy of either row 1 or row 2, then the value of the game for all possible values of p is represented by the lines in the graph in figure 26.4. If Hank plays a mixed strategy, then each possible mixed strategy will generate a new line that represents the value of the game as a function of p. We need not consider Hank's mixed strategies, however, because each of Hank's mixed strategies is a weighted average of the pure strategies. We can therefore deduce that each of Hank's mixed strategies will yield a line that lies between the lines for Hank's pure strategies and contains the point of intersection of the lines for Hank's pure strategies.

How does the graph in figure 26.4 help Dizzy choose a value of p?

Consider several values for p, specifically $p = .1$, $p = .5$, and $p = .8$. Vertical lines through these values of p are shown on the graph in figure 26.5. Notice that each of the vertical lines intersects both line 1 (the value of the game for pure strategy row 1) and line 2 (the value of the game for pure strategy row 2). If Dizzy chooses $p = .1$, then the value of the game can be at a level anywhere on the line $p = .1$ between the points of intersection of $p = .1$ with lines 1 and 2, depending on the strategy that Hank plays. If Hank realizes over time that Dizzy is playing a mixed strategy with $p = .1$, then Hank can move the value of the game up to line 2 by playing a pure strategy of row 2 (guess curve ball). The value of the game on line 2 is given by $.150p + .300(1 - p)$, which equals .285 if $p = .1$. If Dizzy chooses $p = .5$, then Hank can force the value of the game up to the intersection of $p = .5$ and line 1 by playing a pure strategy of row 1 (guess fast ball). The value of the game at that point is given by $.400p + .250(1 - p)$, which equals .325 if $p = .5$. Similarly, Hank can play a pure strategy of row 1 if $p = .8$, yielding a value of the game of .370. The value of the game in each case is the batting average Hank would expect to have under these strategies.

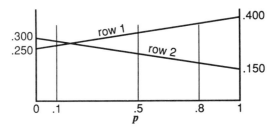

Fig. 26.5. Value of the game for specific values of p

Given that Hank can always force the value of the game to the higher of lines 1 and 2 on any vertical line, the best that Dizzy can do is choose the value of p where lines 1 and 2 intersect. (Recall that Dizzy wishes the value of the game to be as small as possible.) We can estimate this value of p from the graph, but we can obtain an exact value by equating the expressions for each line as follows:

$$.400p + .250(1 - p) = .150p + .300(1 - p)$$
$$.400p + .250 - .250p = .150p + .300 - .300p$$
$$.150p + .250 = -.150p + .300$$
$$.300p = .050$$

Therefore, the value of p at the point of intersection of lines 1 and 2 is $\frac{1}{6}$. This represents the best mixed strategy for Dizzy—randomly mix pitches so that $\frac{1}{6}$ are fast balls and $\frac{5}{6}$ are curve balls. The value of the game at $p = \frac{1}{6}$ is found by substituting $\frac{1}{6}$ for p in the equation for either line, as in

$$.400(1/6) + .250(5/6) = .275.$$

Dizzy can expect Hank to hit .275 against this mix of pitches.

A similar analysis can be performed to find the best mixed strategy for Hank. We should not assume that Hank's mixed strategy will mirror the mixed strategy used by Dizzy. Assign a probability q to row 1 (guessing fast ball) and probability $1 - q$ to row 2 (guessing curve ball). If Dizzy plays a pure strategy of column 1 (throws fast balls), then the value of the game as a function of q is

$$.400q + .150(1 - q).$$

If Dizzy plays a pure strategy of column 2 (throws curve balls), then the value of the game is

$$.250q + .300(1 - q).$$

Graphs of both lines are shown in figure 26.6. Since Dizzy will always attempt to push the value of the game as low as possible, the best that Hank can do is to choose the value of q at the intersection of these lines. Equating the expressions for the two lines yields $q = .5$, so Hank should randomly guess fast ball half the time and curve ball half the time. The value of the game at $q = .5$ is

$$.400(.5) + .150(.5) = .275;$$

so Hank can expect to hit .275 using this guessing strategy.

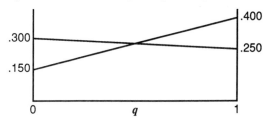

Fig. 26.6. Value of the game for each of Dizzy's pure strategies

The mixed strategies calculated for Dizzy and Hank allow each player to get the best possible payoff from the game over the long run; therefore, these strategies are called *optimal* mixed strategies. Notice that the value of the game is the same with Dizzy's optimal mixed strategy as with Hank's optimal mixed strategy. This is an example of a general result of game theory concerning two-person, zero-sum games: *Optimal mixed strategies exist for both players; furthermore, the value of the game for each player is the same if they each play optimal mixed strategies.*

If one player has more than two strategies but the other player has only two strategies, the situation is slightly more complicated than with 2 × 2 games, as the next example shows.

A Money Game

Jack and Diane agree to play a game in which they each secretly lay down a certain amount of money on each turn. Jack can put down a one-dollar bill or a twenty-dollar bill. Diane has the option of setting down no money, a five-dollar bill, or a ten-dollar bill. If the total of the money set down is even, then Jack must pay Diane that amount. If the total is odd, then Diane must pay Jack that amount. The payoff matrix for this game is given in figure 26.7, with payoffs given from the perspective of the row player Diane. Notice that the game does not have a saddle point, since the maximum of the row minimums is -1 and the minimum of the column maximums is 6.

		Jack	
		$1	$20
	0	-1	20
Diane	$5	6	-25
	$10	-11	30

Fig. 26.7. Payoff matrix for the money game between Jack and Diane

Let us first analyze the game from Jack's perspective, the player who has only two options. Assign a probability q to the chance that Jack plays column 1 ($1) and a probability $1 - q$ to column 2 ($20). The expected payoffs from each of Diane's three pure strategies are as follows:

$$\text{Row 1:} \quad -q + 20(1 - q)$$
$$\text{Row 2:} \quad 6q - 25(1 - q)$$
$$\text{Row 3:} \quad -11q + 30(1 - q)$$

These three linear functions are sketched in figure 26.8 for $0 < q < 1$.

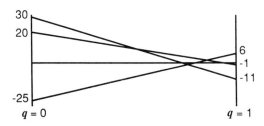

Fig. 26.8. Value of the game for each of Diane's pure strategies

Since Diane wants to push the value of game up as high as possible, the value of the game for a particular q is determined by which of the three lines is located highest in the plane for that q-value. The set of line segments that determine the value of the game are outlined in figure 26.9. Jack wants the value of the game to be as low as possible; therefore, his optimal mixed strategy is determined by the point of intersection of the lines from row 1 and row 2. Solving these two equations simultaneously yields a probability q equal to 45/52. Substituting this q into either the row 1 or row 2 equation results in a value of the game equal to 95/52, or about $1.83.

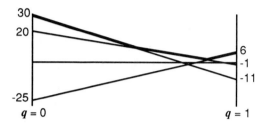

Fig. 26.9. Worst-case value of the game from Jack's perspective

Notice that Diane's row 3 strategy does not come into play in determining Jack's optimal mixed strategy; therefore, Diane should use only rows 1 and 2 in her optimal mixed strategy. Assigning row 1 a probability p and row 2 a probability $1 - p$ leads to the following expected values of the game for each of Jack's pure strategies:

$$\text{Column 1: } -p + 6(1 - p)$$
$$\text{Column 2: } 20p - 25(1 - p)$$

Equating these two expressions and solving for p, which corresponds to finding the intersection of two lines, yields a probability $p = 31/52$. Substituting this p into either equation above gives a value of the game equal to 95/52, as we found previously.

The optimal mixed strategy for Jack is to lay down $1 with a probability of 45/52 and to lay down $20 with a probability of 7/52. This randomization could be accomplished by using a standard deck of 52 cards. At each play of the game, Jack would first draw a card from the shuffled deck. If the card is among the ace through seven of hearts, then Jack would set down $20; otherwise, he would set down $1. The optimal mixed strategy for Diane is to lay down no money with a probability 31/52, to lay down $5 with a probability 21/52, and to never lay down $10.

The value of the game is approximately $1.83, which means that on the average Diane will gain $1.83 on each play of the game. Jack would be wise not to get involved in this game at all!

Any $n \times 2$ or $2 \times n$ zero-sum game can be analyzed using the same technique as that demonstrated in the game between Jack and Diane. Another fact concerning two-person, zero-sum games is relevant to these problems: *The number of nonzero components in each player's optimal mixed strategy is the same for both players.*

Determining optimal mixed strategies in $m \times n$ zero-sum games in which both m and n are greater than 2 is more complicated than the previous examples. Such games can be solved using linear programming and the simplex algorithm, or a method of successive approximations given by Brown's algorithm can be applied. (See Thomas 1986.)

NON-ZERO-SUM GAMES

Not all competitions and conflicts are best formulated as zero-sum games. Many situations exist in which the gain for one player is not equal to the loss for the other player. Such games are called *non-zero-sum games*. Two well-known, non-zero-sum games are "chicken" and "the prisoner's dilemma."

The game of chicken was often played by teenagers in the 1950s. The contest involved two individuals who drove cars toward each other at high speeds. The "chicken" is the one who swerves to avoid the oncoming car. The individual who does not swerve gains esteem in the eyes of the peer group. A form of chicken was played in a famous scene in "Rebel without a Cause," a movie that helped make James Dean a cult hero.

Chicken can be modeled as a two-person game. Each player has two strategies, to swerve or to go straight. We first consider the payoffs for the row player. The row player's best outcome is to go straight when the column player swerves. If neither player swerves, a disaster occurs, and this is the worst outcome for the row player. If the row player swerves and the column player goes straight, this is the second-worst outcome for the row player. Some might claim that this is worse than if both players swerve; however, at least in this case, the row player is alive and can play again in an attempt to recapture lost esteem. The second-best outcome for the row player is when both players swerve. Some prestige is lost, but not nearly as much as when the column player goes straight. The row player's outcomes are ranked from 1 to 4, with 4 the best, and are displayed as the first elements in the ordered pairs shown in the payoff matrix in figure 26.10. The outcomes for the column player also can be ranked, but the roles are reversed. The column player's payoffs are the second elements in the ordered pairs in figure 26.10.

	Swerve	Go straight
Swerve	(3, 3)	(2, 4)
Go straight	(4, 2)	(1, 1)

Fig. 26.10. Payoff matrix for the game of chicken

Chicken is a non-zero-sum game, since the benefit to one player does not result in an equal loss to the other player. Notice the result if both players try to obtain their best payoffs, that is, both try to receive a payoff of 4. Each player will choose a go-straight strategy, leading to disaster, and both players will obtain their worst outcomes. By contrast, if both decide to swerve, then both players can obtain their second-best outcomes.

The game of chicken has much in common with many human conflicts. For example, labor negotiations can be modeled as a game of chicken with labor and management as the two players. Given that each has the choices "compromise" and "don't compromise," the payoff matrix is shown in figure 26.11. The payoff matrix illuminates a primary difficulty with labor-management conflict. Suppose labor is out on strike and both sides have taken a "don't compromise" position. Labor has strong motivation to stay with its position, since if management chooses to compromise, then labor will receive its best outcome. By contrast, management has similar motivation to not compromise. Both sides would be well served if the outcome of the game could be moved from the (1, 1) slot to the (3, 3) slot.

With this objective in mind, suppose labor decides unilaterally to compromise and to return to work. The outcome of the game now moves to the (2, 4) slot. Unfortunately, management now has no motivation to compromise, having achieved its best outcome. The game will be stuck in the (2, 4) payoff unless labor decides to court disaster by calling another strike. Likewise, if management decides to compromise first, then the game will move to the (4, 2) slot, and labor will have no motivation to compromise. Clearly, for the game to move to the (3, 3) slot requires both sides to compromise simultaneously. This assumes a certain amount of trust between the two players—a common sticking point in labor negotiations.

		Management	
		Compromise	Don't compromise
Labor	Compromise	(3, 3)	(2, 4)
	Don't compromise	(4, 2)	(1, 1)

Fig. 26.11. Chicken applied to labor negotiations

The prisoner's dilemma is a game often played out as a "good cop/bad cop" routine in the movies and on television. Two people are arrested with stolen property in their possession, and the police interrogate them in separate rooms. Each criminal knows that the police do not have enough evidence to convict them on robbery charges. If they both keep quiet, then they will receive light sentences for possession of stolen goods. The police, however, offer each of them the same deal. If one will confess and turn in the silent partner, then the one who squeals will go free. The other will receive a stiff sentence for robbery and for not assisting the police. If both end up talking, then they both will serve time for robbery, but less time than if they had not helped the police.

Each player (the criminals) in the prisoner's dilemma has the option of silence or defection. A player's best outcome is to defect while the other remains silent. The worst outcome is to remain silent while the other defects.

A player's second-best outcome is to remain silent while the other remains silent. The second-worst outcome is for both players to defect. The payoff matrix in figure 26.12 shows the outcomes ranked from 1 (worst) to 4 (best) with the row player's payoffs listed first in each ordered pair.

	Silence	Defection
Silence	(3, 3)	(1, 4)
Defection	(4, 1)	(2, 2)

Fig. 26.12. Payoff matrix for the prisoner's dilemma game

The row player's defection payoffs dominate the silence payoffs, and likewise for the column player. Both players are motivated to choose the defection strategy, placing the outcome of the game in the (2, 2) slot. If the players follow the dictates of this model, then the police will get their wish— conviction of both criminals on robbery charges.

The prisoner's dilemma can be used to model arms control between the U.S. and the USSR. Each side has the option to arm or to disarm. The best outcome for one side is to arm while the other side disarms, which is also the worst outcome for the side that disarms. The second-worst outcome is for both sides to arm, thus diverting economic resources from the people in each country. The second-best outcome is for both sides to disarm, thus allowing the economies to flourish in relative security. These outcomes are represented in the payoff matrix in figure 26.13.

		USSR	
		Disarm	Arm
U.S.	Disarm	(3, 3)	(1, 4)
	Arm	(4, 1)	(2, 2)

Fig. 26.13. Payoff matrix for the arms race

For each side, the arm strategy dominates the disarm strategy, thus leading to an arms race, which is represented by the outcome (2, 2). Once the game is in the arm-arm slot, neither side has motivation to disarm unilaterally. To do so would lead to the worst outcome for the side that disarms and hand the best outcome to the side that arms. Suppose the outcome of the game has somehow moved to the disarm-disarm slot, with outcome (3, 3). The payoff matrix reveals that each side is under continuous pressure to arm, which moves the outcome to the (4, 1) or (1, 4) slot. The side that stayed with the disarm strategy is now forced to arm to avoid a worst-case outcome, thus moving the outcome of the game to the (2, 2) slot. This analysis suggests that the disarm-disarm outcome is unstable, eventually degenerating to the arm-arm outcome that, unfortunately, is a stable outcome in the game matrix. These observations from the payoff matrix are not inconsistent with the recent history of arms-control negotiations.

SUMMARY

In this brief survey of game theory, we have seen mathematics applied to competitive situations to which students can relate. The mathematical techniques are not complex, yet the resulting analysis is powerful. Furthermore, the mathematics has been developed mainly within the last fifty years and continues to be applied to new areas of human endeavor, giving students a sense of the dynamic, expanding nature of mathematics. The problems in game theory are open-ended and thought-provoking. In addition, students can use game theory to model situations from their own experiences. In the midst of studying game theory, students will be motivated to practice their skills in manipulating linear functions. Analyzing a payoff matrix, particularly in non-zero-sum games, also forces students to use precise, logical reasoning. All these benefits furnish a strong rationale for including game theory in secondary school mathematics courses.

BIBLIOGRAPHY

Bradley, Ian, and Ronald L. Meek. *Matrices and Society.* Princeton, N.J.: Princeton University Press, 1986.

Consortium for Mathematics and Its Applications. *For All Practical Purposes: Introduction to Contemporary Mathematics.* Edited by Lynn A. Steen. New York: W. H. Freeman & Co., 1988.

Dresher, Melvin. *The Mathematics of Games of Strategy.* New York: Dover Publications, 1981.

Thomas, L. C. *Games, Theory, and Applications.* Chichester, England: Ellis Horwood, 1986.

Williams, John D. *The Compleat Strategyst.* New York: Dover Publications, 1986.

Zagare, Frank C. *Game Theory: Concepts and Applications.* Quantitative Applications in the Social Sciences series, vol. 41. Newbury Park, Calif.: Sage Publications, 1984.

————. *The Mathematics of Conflict.* HiMAP Module #3. Arlington, Mass.: Consortium for Mathematics and Its Applications, 1985.

27

A Knapsack Problem, Critical-Path Analysis, and Expression Trees

Janet H. Caldwell
Francis E. Masat

O VER the last three years, secondary mathematics teachers have been learning about discrete mathematics and applying what they have learned in summer mathematics institutes offered by the South Jersey Mathematics, Computer, and Science Instructional Improvement Project (Mc-Siip) at Glassboro State College. As part of these institutes, teachers and college faculty have developed and field-tested activities appropriate for use at several grade levels.

This chapter describes three classroom activities in various areas of discrete mathematics. The activities focus on the following topics: the knapsack problem, critical-path analysis, and expression trees.

A KNAPSACK PROBLEM:
PACKING FOR A TWO-WEEK HIKE

This activity introduces algebra students to the concept that there may be many solutions to a problem and that we must often search for a best solution among many. The activity also shows how systematic trial-and-error techniques can be used effectively in problem solving.

Problem. You are taking a two-week hike and will be backpacking everything you need. You have made a list of items to take, and your list has each item's weight and its value to you rated from 1 to 5, with 5 being the highest (fig. 27.1). If you can carry only 30 pounds, what should you take along to get the highest number of value points?

The stated problem lends itself to solutions based on weight (see fig. 27.2). Note that items 1 to 4 total 27 pounds, with a point value of 16. If we use items 5 and 7 for our last 3 pounds, the value is $16 + 3 + 5$, or 24. If we use items 6 to 8 for our last 3 pounds, we get $16 + 2 + 5 + 2$, or 25, one point higher. To help the students find other point totals, prepare a worksheet with just the five headings for them to complete.

Item	Weight	Points (Value)
1. Food	14 lbs.	5
2. Sleeping bag	8 lbs.	4
3. Hygiene kit	2.5 lbs.	4
4. Pans	2.5 lbs.	3
5. Clothes	2 lbs.	3
6. Extra shoes	1 lb.	2
7. Backpack	1.5 lbs.	5
8. Camera	0.5 lb.	2

Fig. 27.1

Item	Weight	Points	Cumulative weight	Cumulative points
1. Food	14 lbs.	5	14 lbs.	5
2. Sleeping bag	8 lbs.	4	22 lbs.	9
3. Hygiene kit	2.5 lbs.	4	24.5 lbs.	13
4. Pans	2.5 lbs.	3	27 lbs.	16
5. Clothes	2 lbs.	3	29 lbs.	19
6. Extra shoes	1.5 lbs.	2	30.5 lbs.	21
7. Backpack	1 lb.	5	31.5 lbs.	26
8. Camera	0.5 lb.	2	32 lbs.	28

Fig. 27.2

Another useful approach is to arrange the items in order of value as in figure 27.3. Here, the first five items total 30 pounds, with a total value of 24—not as good as before! To help students find other values using this approach, give them the worksheet with the five headings again.

Item	Weight	Points	Cumulative weight	Cumulative points
1. Food	14 lbs.	5	14 lbs.	5
7. Backpack	1 lb.	5	15 lbs.	10
2. Sleeping bag	8 lbs.	4	23 lbs.	14
3. Hygiene kit	2.5 lbs.	4	25.5 lbs.	18
4. Pans	2.5 lbs.	3	28 lbs.	21
5. Clothes	2 lbs.	3	30 lbs.	24
6. Extra shoes	1.5 lbs.	2	31.5 lbs.	26
8. Camera	0.5 lb.	2	32 lbs.	28

Fig. 27.3

Another interesting approach is to have your students compute the points-per-pound ratio for each item. Doing this, the first item we would choose is the knapsack, since its ratio is 5:1, or 5. Our next item is No. 8, the camera, since its ratio is 2:0.5, or 4. (Our food, with a ratio of 5:14, or 0.36, may well be left out using this scheme!) Which items would your students take next? Another worksheet, with a blank column for the ratios, is useful here.

More problems can be constructed by varying the number of items, representing weights in metric units, and changing the value points. Alternatively, you can ask the students to choose items for a special dinner, replacing weights with cost and setting a cost limit.

CRITICAL-PATH ANALYSIS: A YEARBOOK PRODUCTION SCHEDULE

This activity gives beginning algebra students practical experience in organizing and quantifying information. It also introduces the discrete mathematics concept of critical-path analysis.

Problem. You have been chosen to work on your school's yearbook and have been given a list of tasks to supervise. The order of the tasks and the amount of time for each task are listed in figure 27.4. As you make a schedule for these tasks, you must decide whether or not you can have the yearbook ready for the printers in four months (16 weeks). Can you?

Tasks to be done	Time required	Preceding tasks
A. Decide name and length of yearbook	1 week	None
B. Obtain printing estimates	3 weeks	A
C. Select a printer	1 week	B
D. Put up posters for stories and photos	1 week	A
E. Collect material submitted	5 weeks	D
F. Select material for yearbook	3 weeks	A, E
G. Decide layout	3 weeks	C, F
H. Final review of book	2 weeks	G
I. Send to printer	1 week	H

Fig. 27.4

Distribute the problem and let students think about it before demonstrating some approaches. For example, just adding up all the times gives the impression that 20 weeks are needed to get ready. But, since some tasks can be done at the same time, like B and D, a better estimate (fewer weeks) should be possible.

Since such information is often more easily understood in chart form, it is useful to list the information by the sequence of tasks, that is, which task comes next. You can then ask how long it takes to finish the different combinations of tasks. Depicting the information visually yields the chart in figure 27.5.

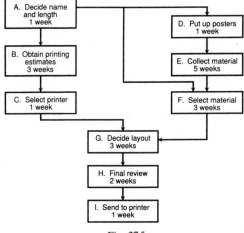

Fig. 27.5

This chart translates to the directed graph in figure 27.6, in which each vertex represents a task and the time required to finish it.

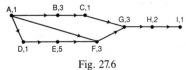

Fig. 27.6

The possible paths, beginning with vertex A,1, and their required total times (sums) can now be read directly from the diagram.

$$A, B, C, G, H, I \text{ or } 1 + 3 + 1 + 3 + 2 + 1 = 11$$

$$A, F, G, H, I \text{ or } 1 + 3 + 3 + 2 + 1 = 10$$

$$A, D, E, F, G, H, I \text{ or } 1 + 1 + 5 + 3 + 3 + 2 + 1 = 16$$

Since enough time must be provided to accomplish *all* the tasks, you can compare these paths and see that the last path listed yields the minimum amount of time needed. That is, the last path is the only one that allows enough time to do *all* the tasks. Therefore, the answer to the original problem is yes, since the student can finish all the required tasks in 16 weeks.

Additional exercises can be constructed easily. For example, change the tasks C, E, and F in figure 27.4 as follows:

C. Select a printer	2 weeks	B
E. Collect material submitted	2 weeks	D
F. Select material for yearbook	2 weeks	A, E

The minimum time needed with these requirements should be 12 weeks.

The solution method demonstrated here is called PERT, for Program Evaluation and Review Technique. It was developed in 1958 for the U.S. Navy, although similar techniques were invented in England, France, and Germany at about the same time. It is currently widely used in business, and many microcomputer programs are available to generate PERT charts to aid in scheduling complex projects. The usefulness of this process for programming computers to schedule and estimate completion times for large projects with hundreds of tasks should be obvious.

EXPRESSION TREES: CALCULATORS AND ORDER OF OPERATIONS

A special kind of graph, called a *tree*, is very useful in computer science and mathematics. One use of trees is to represent and analyze the order of operations.

Problem. Katie and Philip each enter the following sequence of keystrokes on their calculators:

$$4 \boxed{+} 6 \boxed{\times} 7 \boxed{-} 4 \boxed{\div} 2 \boxed{=}$$

Katie gets an answer of 33, but Philip gets 44. Which answer is correct, and why is one calculator wrong?

Expression trees are binary trees. Each operation corresponds to a vertex, and each number is a terminal vertex. The root is the last operation performed in the expression.

First, let us look at Katie's calculator. Her calculator gives the answer after each operation; so after she keys in 4 $\boxed{+}$ 6, the calculator automatically shows 10. The tree in figure 27.7 shows how her calculator works.

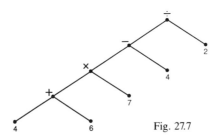

Fig. 27.7

When the graph is read from the bottom, each operation represents a vertex with a right child (which must be a number with this calculator) and a left child (which may be a number or the result of a previous operation). This calculator does not follow the standard order of operations; it simply calculates left to right—

$$((((4 + 6) \times 7) - 4)) \div 2.$$

Philip's calculator, by contrast, gives an answer only when the $\boxed{=}$ key is pushed; it follows the usual order of operations rules. This sequence can thus be represented by the expression tree in figure 27.8. Each operation is represented by a vertex with a right child (which may be a number or an operation) and a left child (which may be a number or an operation).

It is useful for students to consider additional examples of expression trees. (See figs. 27.9 and 27.10.)

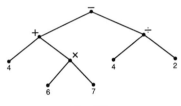

Fig. 27.8

Example 1. The expression tree for $7 + 2 \times 3$ is

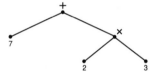

Fig. 27.9

Example 2. The expression tree for $b \times (c - d) + a$ is

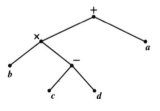

Fig. 27.10

These examples are a good starting point for a second activity that provides two types of exercises. The first group of exercises requires students to draw expression trees that represent given arithmetic or algebraic expressions (similar to the examples). The second group of exercises asks, "Given the following expression tree, what was the original arithmetic or algebraic expression?" Some examples as guides are in figure 27.11.

Trees were first used in 1847 by Gustav Kirchoff in his work on electrical networks. Later, Arthur Cayley used them in his work in chemistry. Today,

(A)

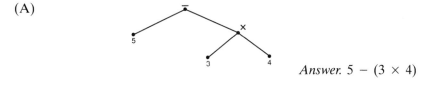

Answer. $5 - (3 \times 4)$

(B)

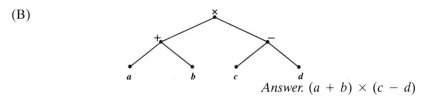

Answer. $(a + b) \times (c - d)$

(C)

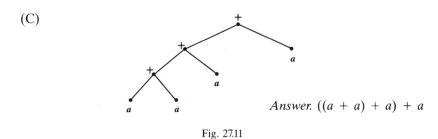

Answer. $((a + a) + a) + a$

Fig. 27.11

trees and associated algorithms are used throughout mathematics and computer science to organize and manipulate data.

SUMMARY

Using these activities and others, teachers in summer mathematics institutes have found ways to include discrete mathematics in the secondary school mathematics curriculum. They have found that such activities are more tangible to their students and make students better able to appreciate the power of mathematics. These activities help students to value mathematics, to communicate mathematically, and to reason mathematically. They allow students to be mathematical problem solvers, using real-life problems. In addition, they show students that mathematics is more than manipulating numbers and symbols; mathematics is a powerful representational tool.

BIBLIOGRAPHY

Consortium for Mathematics and Its Applications (COMAP). *For All Practical Purposes: An Introduction to Contemporary Mathematics.* Edited by Lynn A. Steen. New York: W. H. Freeman & Co., 1988.

Dossey, John A., Albert D. Otto, Lawrence E. Spence, and Charles Vanden Eynden. *Discrete Mathematics.* Glenview, Ill.: Scott, Foresman & Co., 1987.

28

Design Your Own City:
A Discrete Mathematics Project
for High School Students

Carol A. Bouma

THE NCTM *Curriculum and Evaluation Standards for School Mathematics* (1989) calls for the introduction of discrete mathematics topics into the secondary school curriculum that traditionally have not been explored by students below the undergraduate level. The following project description illustrates how some discrete mathematics topics have been investigated successfully by high school students, when taught at an appropriate level of complexity.

At the secondary school level, the *Standards* also recommends variety in modes of instruction and assessment. One recommended mode is student exploration within, and accountability to, small teams of peers whenever the subject matter permits such learning activity. This chapter describes such a learning experience, some student reactions, and a method of evaluation for the shift in the pattern of instruction from teacher-centered learning to cooperative-student-group learning.

BACKGROUND

Piloted in the fall of 1988 at the Park School in Brooklandville, Maryland, this project synthesized topics that students had recently explored for the first time—optimization algorithms, networks and circuit theory, and multitask processing. *For All Practical Purposes: Introduction to Contemporary Mathematics* (COMAP 1988) was the text that served as the basis for the course, supplemented by work taken from the School Mathematics Project (1987) texts used in mathematics courses at the Park School.

THE CHALLENGE

Charged with the responsibility of being "consultants" hired to design the layout and services of a small city, students worked in groups of two or three. Each group had in common only a map of a given amount of land and the challenge to use certain discrete mathematics topics to prepare a convincing design proposal. All the general and specific requirements for the project were spelled out for the students on a one-page specification sheet including a time frame for accomplishing the project through drafts and revisions.

BEGINNING THE WORK

"Wouldn't you use a Hamiltonian circuit instead of an Euler circuit for a monorail system carrying visitors . . . ?"

"When we studied critical-path analysis, I thought I understood it, but when I try to apply it to my construction schedule, I'm wondering about. . . ."

"You can't use bin-packing for tollbooths!" "Why not? Some methods would smooth out traffic flow over bridges. . . ."

"My equations should give me a maximum point!"

These were the kinds of comments I heard students making to each other in the early stages of the project, as they energetically developed, dissected, and defended their creative insights with the discrete mathematics techniques recently introduced.

The energy with which each group attacked such a complex problem was only partially explained by the real control they had over the scope of their effort. The rest of the excitement and contagious motivation seemed to come from—

- the chance to create a piece of the world "as it should be" (by their own definition);
- the natural joy of contributing to, and being responsible for, the work of a small group of friends;
- the pride in achieving a worthwhile product of prolonged effort and thought.

Although the students had studied, practiced, and used each skill to solve problems in class and in homework assignments, they found that a new and richer understanding was required to actually use the skills together to optimize plausible situations. List-processing algorithms, for instance, were no longer something you memorized for a test, but something you molded, squeezed, stretched, and spent the weekend massaging into a form you could reasonably apply to solve your construction problem. Needing to choose the best bin-packing algorithm for loading your city's school buses made researching the various forms and their distinguishing features a necessary, logical step toward a higher goal. All the management-science techniques that were required elements in the project took on a new value, as students began to use them as their own mental tools for accomplishing physical tasks.

THE STUDENTS

My students were college-bound seniors who had elected this Contemporary Mathematics Seminar and had varying levels of interest in pursuing mathematics as undergraduates. All the students had successfully completed the prerequisite mathematics curriculum, and some had had an introduction to calculus. Although their curricular backgrounds were similar, there was real diversity in the students' facility with, goals for, and concept of the mathematics they had already studied. Some were intrigued by the excursions into pure mathematics in earlier courses and were averse to anything so "applied" as the discrete mathematics topics we had just investigated. Initial resistance to the effort and the requirements of the project was greatest among these theoretically inclined members of the class. However, some of these same students eventually felt most rewarded by the work. The open-ended nature of the project gave the students the opportunity not just to express but to revel in this diversity of interest, for they were all accomplishing the same broadly defined goals, using their own unique talents and perspectives:

- Artists rendered street layouts, postal routes, and mining-tunnel diagrams with great attention to visual detail.
- Social science devotees experimented with the integration of different economic groups in housing and employment opportunities.
- Ecology enthusiasts studying the effects of nuclear energy in other classes structured their city around a nuclear power plant and envisioned the constraints and advantages in mathematical terms.
- Mathematically sophisticated students, taking pleasure in their ability to apply the techniques in novel ways, divided their city geographically so that each team member could apply all the required techniques to a particular setting within the group's city—mining, farming/trucking, or housing subsections.

Most groups divided up the effort, with individuals choosing techniques that they felt most comfortable developing, and later reassembled their individual contributions and reviewed the whole proposal. From the same specifications, student groups designed a college town; a city with farming, mining, and housing industries; a town centered on a nuclear power plant;

and a theme park–based resort area. *All* the groups went beyond the required minimum effort as their growing confidence in the work began to yield reasonable, rewarding results.

MATHEMATICAL HIGHLIGHTS: EXAMPLES FROM STUDENT PROJECTS

Following are brief descriptions of the mathematical techniques as the students studied them and excerpts of actual student proposals illustrating how the techniques were creatively applied:

Euler and Hamiltonian Circuits

An *Euler circuit* is a tour through a network in such a way that each edge (connecting line segment) is traversed once and only once.

A *Hamiltonian circuit* is a tour through a network that visits each vertex once and only once.

A farming community

When plowing, planting, harvesting, or inspecting is done, Euler circuits should be used. Many farmers would cover the entire length of each row [as shown in fig. 28.1], which gets it all in one pass, but would then have to drive the tractor . . . back to the beginning. . . . If, however, the farmer used an Euler circuit, it would look like [fig. 28.2], which eliminates . . . an extra four thousand feet each time he works. That doesn't seem like much, but three [fields] a day, two hundred times a year is 12 000 000 feet saved every five years. Vehicles will last longer and farmers will save time from less maintenance.

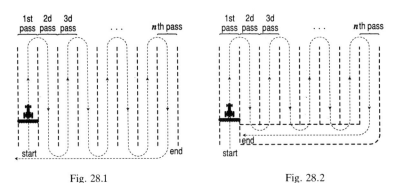

Fig. 28.1   Fig. 28.2

A college town

In order for prospective students to see all the critical areas of the campus [on a tour] in as little time as possible, a Hamiltonian circuit has been constructed. Leaving the Administration Building, students visit the main library, and continue on to. . . .

A resort town

Through our studies, the development office found that the most efficient tour for the resort was a Hamiltonian circuit. . . . The operation of its monorail system . . . connects all of the resorts . . . [which] represent the necessary vertices, and visits each vertex once and only once within the circuit.

Critical Paths

Critical paths are analyses of schedules of tasks being performed by multiple processors to determine those tasks that, by virtue of the prescribed order of their performance, define the minimum completion time of the whole job. Shortening the time required for these critical tasks may thus shorten the time that all resources need to commit to the effort. See figure 28.3.

City Construction . . .

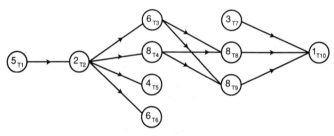

Circled numbers are in months.

T1: Landscaping (leveling hills . . . levees, wildlife preserve)

T2: Underground systems (sewer, electrical, telephone)

T3: Train tracks

.

.

.

T10: Municipal amenities (fountains, parks, etc.)

Critical Paths: T1, T2, T4, T8, T10 and T1, T2, T4, T9, T10

Fig. 28.3

Task-scheduling Algorithms

Task-scheduling algorithms are methods of assigning tasks to the processors that will perform them, with the goal of minimizing the total time that the processors are idle.

List-processing algorithms rank tasks according to their relative priorities (preassigned, based on the nature of the work itself). Their readiness for processing is determined by a directed graph, which shows the time required for each task and those tasks that must be completed before a given task can be begun. Ready tasks are assigned to the next free (idle) processor in order of priority.

Decreasing-time-list algorithms rank tasks according to the length of time each requires, with the longest task first. Ready tasks are assigned to the next free processor according to this ranking under the constraints of a directed graph. Figure 28.4 is an example of such a list.

Construction Tasks (ranked by length)

T1	Industrial center	1.50 years
T2	Town center	1.00 year
T3	Regular housing	0.75 year
T4	Highways	0.50 year
T5	High-priced housing	0.50 year
T6	Secondary roads	0.50 year
T7	Power plant	0.50 year
T8	Electric lines	0.25 year

Fig. 28.4

In both these methods, the directed graph, known as an *order-require-ment digraph,* states which tasks must precede others, regardless of ranking. In the student's exa..ple shown in figure 28.5, highways, T4, had to be completed before work on the industrial center, T1, could begin, even though this latter task was higher on the priority list.

Order Requirements

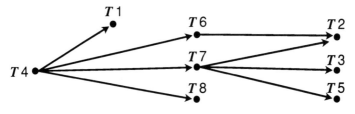

Fig. 28.5

A town built around a nuclear power facility

[Applying a] decreasing-time-list algorithm [to an order-requirement graph and priority list of construction tasks as in figs. 28.4 and 28.5] results in the construction schedule in figure 28.6.

Bin-packing Algorithms

Bin-packing algorithms are methods of minimizing wasted space when packing indivisible units of varying lengths and volume into equal-capacity chambers or bins. (These methods may apply to time or to other more abstract resources.)

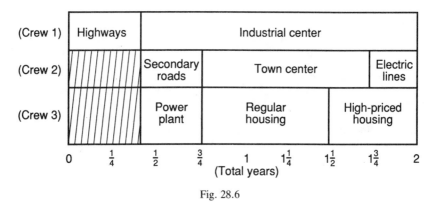

Fig. 28.6

The *next-fit method* packs one bin until the next unit will not fit. Close this bin and open a new one.

The *first-fit method* keeps partially filled bins open. Pack the next unit in the first open bin in which it will fit. Open a new bin if no previously opened bins have enough room.

The *best-fit method,* contrary to the usual interpretation of best-fit packing as the method that leaves the least amount of empty space, packs the next unit in the bin that has the most room. If it does not fit, open a new bin.

Next-fit, first-fit, and *best-fit decreasing methods* sort the units by length before packing, using the given algorithm.

A mining town

The first algorithmically relevant stop [on our tour] is right outside the mines, where we incorporate the mathematical wonders of bin-packing (named, of course, after the famous "algorithmatician" Binjamin Packing). . . . The ore is already encased in packages when it is brought out. . . . Where does bin-packing come in? Well, let's look at the following facts:

- Each case of iron ore weighs 8 kg.
- Each case of nickel weighs 2 kg.
- Each case of zinc weighs 5 kg.

The railroad cars in which the ore will be carried can bear only 21 kg each safely. So the question becomes, What is the fewest [number of] . . . railroad cars in which we can ship the ore?

A town on an expressway

The greatest number of people that will be at the tolls at any one time will be estimated. Then the smallest number of tollbooths will be calculated using the best-fit method to maximize traffic flow. (We use best-fit because people will always choose the shortest line.)

A parking structure

Through our monthly data, we discovered that an average of twenty buses and sixty cars need parking daily. . . . Here the "bins" are the rows for parking, and the items that are "packed" in these bins are the cars and buses. . . . Each bus takes up 3 parking spaces. . . . The developers initiated the study by stating that each bin contained 15 parking spaces, and each floor contained 6 bins. . . . We want to find the minimum number of bins for the cars and buses. The simplest approach is to place the cars and buses in the rows as they enter, . . . filling the first bin and then moving on to the second bin until all the cars and buses are parked. However, this algorithm (next-fit) might not optimize our situation. . . . Furthermore, we realized that we could not use the best-fit decreasing, next-fit decreasing, or first-fit decreasing algorithms because they packed the largest items first, and the buses and cars enter randomly. . . .

Linear Programming

Linear programming is a method of determining optimum mixtures of elements whose availability is variable and subject to constraints that may be expressed and graphed as linear inequalities. The criterion for determining the *optimum* combination or mixture (e.g., highest profit, greatest number of passengers served, etc.) is expressed as a function of these elements and tested against those mixtures determined to be feasible (i.e., those in the intersection of the solution sets representing constraining inequalities).

Shipping farm produce

The number of trucks needed will be calculated by linear programming. Each year, the farmers produce a minimum of 900 bushels and it has been known to go as high as 1560 bushels. A maximum amount of $9200 has been allocated for shipping. Two types of trucks can be used—a small truck that can carry up to 30 bushels and costs $250 or a large truck that can carry up to 80 bushels and costs $400. A minimum of 25 trucks [will be needed to keep drivers employed]. (See fig. 28.7.)

If S = the number of small trucks and L = the number of large trucks,

$$30S + 80L \geq 900 \quad \text{[minimum load constraint]}$$
$$30S + 80L \leq 1560 \quad \text{[maximum load constraint]}$$
$$S + L \geq 25 \quad \text{[minimum total number of trucks]}$$
$$250S + 400L \leq 9200 \quad \text{[cost constraint]}$$

The shaded region in figure 28.7 shows the set of points (representing mixtures of large and small trucks) that satisfies all the constraining inequalities. This is the *feasible region*, and one of its corner points will yield the optimum cost per bushel.

It turns out that in order to maximize crop exports [for the minimum cost for a bushel], 10 small trucks and 15 large trucks should be used. . . .

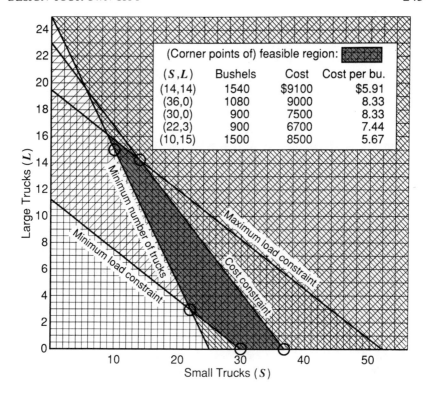

(Corner points of) feasible region:			
(S,L)	Bushels	Cost	Cost per bu.
(14,14)	1540	$9100	$5.91
(36,0)	1080	9000	8.33
(30,0)	900	7500	8.33
(22,3)	900	6700	7.44
(10,15)	1500	8500	5.67

Fig. 28.7

ADAPTING FOR OTHER CLASSES

Project specifications can easily be adapted to suit groups of students with varying goals, abilities, or mathematics backgrounds. Some possible adaptations for other classes include the following:

• Preassigning groups

My students chose their own partners, but teachers may achieve a balance of abilities and work habits by grouping students in advance.

• Limiting the search

Spelling out the specific types of list-processing and bin-packing algorithms that should be considered before a final choice is made may simplify the research/review task for students. They can then focus their efforts on evaluating the relative merits of each method for their problem setting.

• Subdividing the task

Teachers of younger students may need to break down the proposal to require only one application at a time and compile these in group folders as the work progresses.

- Illustrating expectations

Examples of the application of a given technique or algorithm, showing the level of detail expected, can be displayed on an overhead projector or in a handout.

(*Note:* Regardless of the student group targeted, student creativity will be encouraged and sustained only if the teacher deliberately refrains from offering too much guidance in the choices students make initially and in their struggle to adapt the algorithms to those choices. In preparing the first draft, my students found intrinsic motivation in knowing they would be able to surprise the class (and me) with a unique, innovative interpretation of the requirements. Although we did devote some class time to phases of the project where group interactions were important and I was available to answer questions during those sessions, most of the guidance I gave my students, after the initial specification and checking of requirements, took the form of written comments on their rough drafts, charts, and revisions. See the evaluation section below.)

- Restating the objective

Even highly motivated mathematics students initially may be overwhelmed by the prospect of dealing with subjects so foreign to them as building requirements, construction schedules, and so on. It may be necessary to emphasize periodically the prescription to "be more concerned with the choice, structure, and use of the underlying algorithms than with the accuracy of the time estimates, reasonableness of building constraints, and industrial needs." This caution is also appropriate for those student groups who may be preoccupied with the creative aspects of their city design to the detriment of the mathematics required to support their decisions.

EVALUATING PROJECTS

Written comments about drafts on a specially designed group evaluation sheet allowed me to model the way the proposal might be seen from an investor's perspective. I was able to address all members of the group at once, to give the repeated expectations of the "client" in a quantifiable form that would help the students see their own efforts more objectively, and to have a hard copy of my remarks to simplify the evaluation of the final proposal.

Designed as a review, synthesis, and assessment of understanding, the project was intended as a measure of learning that gives those students willing to put in the effort an opportunity to show their understanding in a less stressful setting than a unit test. I found it helpful to weight each intermediate grade more heavily than all that had gone before:

- Outline   30 points
- First draft   50 points
- Last draft/Final proposal   120 points

In this way, students could see progressive improvements in effort and understanding increasingly rewarded. (This need not be announced at the outset, when so much is being assimilated.) The level of real improvement that occurred in my class between the last draft and the final proposal was significant and rewarding. This took place not so much in the grammar or artwork of the presentation but in a new depth of understanding of the mathematics itself. Students who had paid little attention to the choice or development of algorithms earlier saw that that lack seriously undercut the convincing nature of their proposal (and their grade!). When challenged to defend their assumptions, some groups were forced to reexamine the appropriateness of the algorithms they had chosen and the accuracy and consistency with which they were applied. Even in the final stage of the project, for some students, valuable content objectives were still being realized.

CONCLUSION

Even more than the high quality of their written proposals and the recurring references to the project as a highlight of the course in student evaluations, I was impressed with the level of engagement, ownership, and satisfaction in the work that I saw in the class as a whole. My students not only learned something about the content of twentieth-century mathematics, its value in solving real problems, and its limitations and advantages as a way of looking at the world but also learned something about themselves and about their personal power to bring order to that world, as thinking individuals with persuasive mathematical tools.

REFERENCES

Consortium for Mathematics and Its Applications (COMAP). *For All Practical Purposes: Introduction to Contemporary Mathematics.* Edited by Lynn A. Steen. New York: W. H. Freeman & Co., 1988.

School Mathematics Project. Cambridge: At the University Press, 1987.

Index